ALSO BY HOWARD ENSIGN EVANS

Cache la Poudre: The Natural History of a Rocky Mountain River
(with Mary Alice Evans) (1991)

The Natural History and Behavior of North American Beewolves
(with K. M. O'Neill) (1988)

The Pleasures of Entomology:
Portraits of Insects and the People Who Study Them (1985)

Insect Biology: A Textbook of Entomology (1984)

Australia: A Natural History (with Mary Alice Evans) (1983)

The Biology of Social Insects
(coedited with M. D. Breed and C. D. Michener) (1982)

William Morton Wheeler, Biologist
(with Mary Alice Evans) (1970)

The Wasps (with Mary Jane West Eberhard) (1970)

Life on a Little-known Planet (1968)

The Comparative Ethology and Evolution of the Sand Wasps (1966)

Wasp Farm (1963)

Studies on the Comparative Ethology of Digger Wasps
of the Genus Bembix (1957)

Pioneer Naturalists

The Discovery and Naming of
North American Plants and Animals

Howard Ensign Evans
Drawings by Michael G. Kippenhan

x

JM

A John Macrae Book

HENRY HOLT AND COMPANY   NEW YORK

Henry Holt and Company, Inc.
Publishers since 1866
115 West 18th Street
New York, New York 10011

Henry Holt® is a registered trademark
of Henry Holt and Company, Inc.

Library of Congress Cataloging-in-Publication Data
Evans, Howard Ensign
Pioneer naturalists: the discovery and naming of North American plants and
animals/Howard Ensign Evans; drawings by Michael G. Kippenhan.—1st ed.
p.   cm.
Includes bibliographical references and index.
1. Naturalists—North America—Biography.   2. Botany—North America—
History.   3. Zoology—North America—History.   4. Botany—North America—
Nomenclature—History.   5. Zoology—North America—Nomenclature—History.
6. Plant names, Popular—North America—History.   7. Zoology—North America—
Nomenclature (Popular)—History.
I. Title
QH26.E77   1993
508.7'014—dc20   93-12501
                                                                                                              CIP
ISBN 0-8050-2337-2
ISBN 0-8050-2339-9 (An Owl Book: pbk.)

Henry Holt books are available for special
promotions and premiums. For details contact:
Director, Special Markets.

First published in hardcover in 1993
by Henry Holt and Company, Inc.

First Owl Book Edition—1994

Designed by Paula R. Szafranski

Printed in the United States of America
All first editions are printed on acid-free paper.∞

1 3 5 7 9 10 8 6 4 2
3 5 7 9 10 8 6 4 2
(pbk.)

Contents

The universe is a more amazing puzzle than ever, as you glance along this bewildering series of animated forms—the hazy butterflies, the carved shells, the birds, beasts, fishes, insects, snakes, and the upheaving principle of life everywhere incipient, in the very rock aping organized forms. Not a form so grotesque, so savage, nor so beautiful but is an expression of some property inherent in man the observer—an occult relation between the very scorpions and man. I feel the centipede in me—cayman, carp, eagle, and fox. I am moved by strange sympathies; I say continually "I will be a naturalist."

—RALPH WALDO EMERSON, 1833

Introduction: The Beginnings of Natural History in America

*T*he discovery of a New World in the late fifteenth century sent a shock of wonderment through Europe. It was a shock we in the twentieth century can scarcely appreciate, since our knowledge now extends to distant galaxies. We have traveled to the moon, but we have known the moon was there for a long time. (In any case there was nothing there.) The New World was a total surprise to Europeans: a rich and varied land filled with living things from the Arctic deep into tropical forests. There was nothing in their theology to prepare them for the discovery of a new continent. Had the strange animals been on the Ark, and if so how had they gotten across such a broad ocean? How did it happen that the heavenly sphere also covered the New World, inhabited as it was only by heathen savages?

Men being what they are, they looked upon the New World primarily as a source of riches: gold, silver, spices, novel foods, drugs, and medicines. But this was the time of the Renaissance, and there was much curiosity simply to learn what lay across the ocean. The natural history of America had its crude beginnings with Columbus himself, who said of Hispaniola that its mountains are "lofty [and] most beautiful. . . . And the nightingale was singing, and other birds of a thousand kinds." Of course

Columbus did not hear a nightingale on Hispaniola; very probably he heard the ethereal song of the rufous-throated solitaire that still sings there in the much diminished forests.

It was a young acquaintance of Columbus, in youth a page in the court of Ferdinand and Isabella, who sent back the first report on the wonderful flora and fauna of the New World. Gonzalo Fernández de Oviedo had been appointed overseer of mines on Hispaniola; he was later made governor of Santo Domingo. He was well versed in the classic *Historia Naturalis* of Roman encyclopedist Pliny, and he sought to extend it as he traveled about the West Indies and Central America. His 1535 *Natural History of the West Indies* was widely read in Europe. In it he described the alligator, the armadillo, the manatee, the sloth, the tapir, and other animals strange to Europeans. Among the indigenous plants he described were maize, cassava, papayas, pineapples, and prickly pears. Of course his descriptions were superficial and his accounts not without errors, but for more than a century his book served as a guide for those who ventured across the ocean or who dreamed of such a voyage.

Among those who read Oviedo's *Natural History* was Sir Walter Raleigh, who saw to it that his short-lived settlement on Roanoke Island, Virginia, begun in 1585, contained an eminent scholar. This was Thomas Hariot (1560–1621), who had tutored Raleigh in mathematics and, according to some, led him into the heresies that eventually cost him his head. Hariot remained in Virginia only a few months, but upon his return to England he published A *Briefe and True Report of the New Found Land of Virginia* (1588). He described twenty-eight kinds of mammals, including the skunk, black bear, and opossum. He listed eighty-six birds, including that quintessential American bird, the turkey. Among the trees he described were chestnut, hickory, beech, witch hazel, and sassafras.

With Hariot was John White, a painter who produced stylized but generally recognizable portraits of birds, including the bald eagle, redwinged blackbird, and redheaded woodpecker. He also painted Native Americans in their traditional dress but in stiff, classical poses. Hariot told of the Indians' use of tobacco, the effects of which were "rare and wonderful." Raleigh was soon to introduce tobacco to the court of Queen Elizabeth.

Tobacco was only one of many products that flowed from the New World to the Old. Potatoes, Indian corn (maize), and chocolate (from cacao beans) were to become major items in the diet of Europeans. A less welcome import was the syphilis spirochete, causative agent of the "French disease" of the Italians and the "Italian disease" of the French, but in reality a disease of Native Americans that was introduced to Europe apparently by Columbus's sailors.

Reports from the New World often stretched credulity—hummingbirds and "snakes that rattle" were especially hard to believe. Ralphe Hamor wrote from Jamestown of pigeons "so thicke thet even they have shadowed the skie from us." Nicolas Denyse wrote from Nova Scotia of the "pennogoin," a bird with only stubs of wings "which it beat upon the water to aid in fleeing." Alas, we can learn of the passenger pigeon and the great auk only from the tales of early explorers.

Members of the Royal Society of London, founded in 1662, took great interest in America and often assisted those who visited or settled in the colonies, provided they sent back seeds for English gardens and specimens for their museums. One such settler was John Banister (1650–1692), a young clergyman who came to Virginia and roamed the forests for plants and insects. Here he found "a new world of plants, so strange and monstrouse that I am affraide that they may be thought chameras to be found nowhere but in the brain that drew them." In fact Banister was an excellent artist, and some of his drawings are still preserved in the archives of the British Museum. He sent back to English patrons specimens of some 340 species of plants, 100 insects, and 20 mollusks. Banister never published anything himself, but his notes and drawings were made use of by many others, including the great naturalist-philosopher John Ray, just then preparing his classic *Historia Generalis Plantarum*. Banister was planning to prepare his own natural history of Virginia when he slipped on the rocks at the falls of the Roanoke River and was killed.

These were dangerous times to be botanizing in the wilds. John Lawson (?1670–1711), who arrived from England a few years after Banister's death, was burned at the stake by Tuscarora Indians. Lawson had been appointed surveyor general of the Carolinas, an assignment that took him through the interior of those colonies, still thinly settled by Europeans. He kept a journal of Native American customs and languages, with notes on the flora and fauna. In 1708 he returned briefly to London to oversee the publication of his book *A New Voyage to Carolina*. This book has been reprinted several times, most recently by the University of North Carolina Press (1967), along with some of John White's portraits of Native Americans. The natural history of his *New Voyage* is a curious mixture of fact and folklore. Lawson was keen enough to know that the opossum has a pouch in which the young are carried for a time, but he believed that bats might be the result of a cross between a bird and a mouse. In all, Lawson listed 27 mammals, over 100 birds, and 60 fishes (including whales and dolphins). His list of "insects" includes snakes, lizards, and frogs, but no true insects. Altogether Lawson's book provides an excellent introduction to the Carolinas at the beginning of the eighteenth century as well as a perspective on the primitive state of natural history at the time.

One who enjoyed a better relationship with the Native Americans and a keener eye for nature was Mark Catesby (?1679–1749). Catesby arrived in Virginia in 1712 and gathered seeds and plants for English gardens for several years before returning to England in 1719. He was then approached by several members of the Royal Society, who arranged for him to travel to South Carolina with the newly appointed governor, with a salary of twenty pounds a year for acting on behalf of "the uses and purposes of the Society." Catesby soon began exploring the interior of the Carolinas and Georgia, traveling alone in the wilds and often living with native tribes for short periods. Besides supplying his English patrons with specimens, he collected for himself and prepared watercolor paintings of plants and animals before they lost their true coloration. Catesby left the Carolinas after three years, spending several months in the Bahamas before returning to England. There he began preparation of his *Natural History of Carolina, Florida, and the Bahama Islands*, which in its final form included over 200 brilliantly colored plates depicting 171 plants, 113 birds, 9 mammals, 33 amphibians and reptiles, 46 fishes, and 31 insects. Many of our common names for birds were established by Catesby, for example, bluebird, blue jay, hairy woodpecker, and others. His scientific names were Latin phrases supplied by English zoologists and botanists; they would soon be superseded by the simpler, binomial system of naming plants and animals invented by Linnaeus.

Catesby's book was enormously successful, and American natural history was off to a spectacular start. There were over one hundred and fifty subscribers, including the Prince of Wales and the Queen of Sweden. Catesby was elected a fellow of the Royal Society. Although his work was completed before Linnaeus's major publications, he is honored in the Linnaean name of a well-known amphibian, the bullfrog, *Rana catesbeiana*, and that of the southern red lily, *Lilium catesbaei*.

America's first native-born naturalist, John Bartram (1699–1777), maintained close ties with England throughout his life. A fellow Quaker, London merchant Peter Collinson, did much to encourage him, and Bartram and Collinson developed a close personal friendship, by mail that took many weeks to cross the ocean. Bartram supplied seeds and plants for English gardens, for which he was at first paid twenty guineas a year, enough to allow him to spend his summers roaming about the colonies. Later King George III granted him an annual pension of fifty pounds and the title "King's Botanist." Specimens sent overseas were sometimes devoured by rodents aboard ship or doused with salt water. But many arrived safely. A leaf of the remarkable insectivorous plant, the Venus flytrap, even reached Linnaeus. "I never met with so wonderful a phenomenon," wrote Linnaeus, to whom Bartram was "the greatest natural botanist in the world."

Bartram had a farm on the Schuylkill River near Philadelphia. It was a working farm, but on part of it he planted wild seeds he had collected and even experimented with crossbreeding. He had little formal education, but he devoured all the books he could find on natural history subjects. Through Collinson, Bartram published several articles in the *Philosophical Transactions* (London), on subjects as diverse as digger wasps, snakes' teeth, and mollusks. John Bartram and Benjamin Franklin were good friends and with others founded the American Philosophical Society in 1744. However, it languished for several years until finally revived in 1767. Many of Franklin's experiments with electricity were performed at Bartram's farm, and it was Collinson who saw to the publication of Franklin's *Experiments and Observations on Electricity*, which soon was translated into several languages.

On his travels through the colonies, John Bartram was sometimes accompanied by his son William (1739–1823). Together they discovered over two hundred plants that were later described by European botanists. In October 1765, as they explored swampy country along the Altamaha River in Georgia, they found "several very curious shrubs." In April 1773, William revisited this site. Here he found one of them in bloom as well as in fruit. It was a beautiful small tree, the blossoms having five large, white petals surrounding a clump of orange stamens. Bartram proposed naming it after his father's friend Benjamin Franklin, and his cousin, Philadelphia nurseryman Humphrey Marshall, formally published the name *Franklinia*. The tree proved to grow well in Pennsylvania, and nurseries were soon propagating it for distribution to gardens and parks. But it was never seen again in the wild after 1803. Fossils are known from Eurasia, so it seems probable that *Franklinia* was once widespread and that the few trees that survived in a Georgia swamp were the last of their kind in the wild.

As John Bartram grew older, his sons John, Jr., and William assumed care of his farm and gardens. William had become well known from his book *Travels*, which went through several editions and was translated into several languages. Thomas Carlyle praised it, and Samuel Taylor Coleridge and William Wordsworth borrowed images of wild places from it for some of their best-known poems. Bartram's Gardens became a mecca for naturalists; they are still preserved as part of Philadelphia's park system.

The Linnaean revolution in the naming of plants and animals occurred during the lifetimes of Bartram father and son, but neither was consumed by the passion of the time to attach their own names to as many plants and animals as possible. To be sure, William's *Travels* is filled with scientific names. He returned from his major trip through the Southeast in 1777, but publication of his book was delayed until 1791; in the meantime many of the novelties he had sent to Europe had been named by others. No authorities are cited for the names used in his book, and some in fact are original. However, modern authorities do not accept most of his names, even though they are expressed as Linnaean binomials, since his descriptions tend to be vague. One exception is the warmouth, or great yellow bream, a member of the sunfish family, which Bartram called *Cyprinus coronarius* and portrayed in color with accuracy.

"A very bold and ravenous fish," Bartram reported. The sketches and paintings William Bartram made on his travels were sent to English patrons, and most eventually ended up in the British Museum. Many are quite beautiful, and in 1968 they were reproduced in a lavish volume edited by botanical historian Joseph Ewan and published by the American Philosophical Society.

Although the Bartrams are not credited with formally naming many plants or animals, they are honored in the names *Bartramia* (for the upland sandpiper, named for John by French naturalist René Primevère Lesson) and *Bartramia* (a moss, named for William by German bryologist Johannes Hedwig). (Use of the same name for a plant and an animal is acceptable, since there is little chance of confusion.)

One of Linnaeus's own students, Peter Kalm (1716–1779), arrived in Philadelphia in 1748 and made haste to become acquainted with Benjamin Franklin and John Bartram. He traveled about the Northeast, even to Quebec, wandering through forests filled with mosquitoes, snakes, and Native Americans. He carried several hundred plants back to Linnaeus and, like Catesby and like William Bartram some years later, wrote a book that became widely read: *Travels in North America.* Although filled with novel discoveries, it contained a certain amount of misinformation, and Franklin warned of "strangers that keep journals." Linnaeus named one of America's most beautiful shrubs for him: mountain laurel, *Kalmia latifolia.*

André Michaux (1746–1802), an emissary of French horti-culturalists, arrived in 1785. He was already an eminent botanist, having studied with de Jussieu and Lamarck and traveled through Asia searching for plants. Accompanied by his son François, and with Jefferson's blessing, Michaux spent eleven years scouring the country. After many adventures with storms, Native Americans, and near-starvation, he sailed back to France in 1796. Michaux's book was of a more technical nature than those of Catesby, Kalm, and Bartram: *Flora Boreali-Americana*. This was the first guide to the plants of North America and included descriptions of many new species. After he returned to France, Michaux went to Madagascar in a further search for plants. He died there of fever at the age of fifty-six, but his son François carried on, publishing a book on trees, *North American Sylva*, in 1810. Among other plants, water-primrose, *Jussiaea michauxiana*, is named for the elder Michaux (as well as for his mentor, de Jussieu), while the swamp white oak, *Quercus michauxii*, is named for François.

Other botanists who came from abroad in the eighteenth century included John Clayton, for whom Linnaeus named the spring beauty (*Claytonia*), and Alexander Garden, for whom *Gardenia* was named. Both had planned to settle in the American colonies, but Garden failed to make a strong stand for the rebels, and after the Revolution he was banished to England.

Visits of European naturalists to America continued well into the nineteenth century. In 1832 Prince Alexander Philipp Maximilian (1782–1867) arrived in Boston from Germany, accompanied by a twenty-three-year-old Swiss painter, Karl Bodmer. With assistance from the American Fur Company, the prince and his retinue reached nearly to the headwaters of the Missouri River. Maximilian's *Travels in the Interior of North America* contains observations on natural history as well as on the Native Americans they met. Maximilian's bat (*Centronycteris maximiliani*) is named for him.

Bodmer's paintings have become classics. The Native Americans were portrayed in rich colors as noble and beautiful people, and western landscapes as richly exotic. Many of those he painted were dead of smallpox within a few years, and the landscapes are now greatly transformed.

There were other Europeans who came to the New World to study its fauna and flora, and well into the nineteenth century there were Americans who looked to Europe for guidance and for the identification of plants and animals. But to a degree the American Declaration of Independence was also a declaration of scientific independence. Its author, Thomas Jefferson, was a man of strong scientific bent who was eager to expand knowledge of the continent all the way to the Pacific. He encouraged Michaux, who got as far as the Mississippi, and planned the expedition of Lewis and Clark (1804–1806), so enormously productive in discoveries. Many of the specimens taken by Lewis and Clark were named and described by Americans such as George Ord and Alexander Wilson, and many found their way to Peale's Museum in Philadelphia rather than to European museums. But of course ties with Europe were never broken completely, science being by definition international in scope.

To naturalists of the eighteenth and nineteenth centuries the discovery, description, and naming of new plants and animals was all-important. Not only did they feel they were performing a major service (as they were), but it was a source of self-gratification—the recognition associated with their discoveries. While discovery and naming are only the first steps toward knowledge, they are important first steps; one must know an organism exists, then have a name as a handle with which to discuss it. The original natives of America had names for most of the plants and animals in their environment; after all, they needed to recognize those that were good to eat, those that were useful to cure illnesses, those that were dangerous or poisonous.

When Europeans arrived in the New World, one of their first requirements was to find names for the plants and animals they encountered. Often they used names similar to those in Europe, even though that sometimes resulted in misnomers—robin for a bird not closely related to the English robin redbreast, warbler for a group of birds quite different from European warblers. At other times they used Native American names (tobacco, maize, moose) or descriptive names such as skunk cabbage or whip-poor-will. Often there were several names for the same organism. Common mullein, for example, was known in various parts of the British Isles as hightaper, woollens, Jupiter's staff, and hare's beard; after its introduction into America it was sometimes called flannel plant or, in Quebec, tabac du diable. Furthermore, German, French, Spanish, and other languages all had their own names for the same plant or animal. As discoveries throughout the world added steadily to the objects to be named, it became more and more difficult to know to what one was referring.

The first steps in organizing this morass of names were undertaken by European naturalists such as John Ray, who in the early eighteenth century used dead languages, Latin and Greek, to avoid national and regional differences in names. Each organism was called by a Latinized phrase, of which the first word was the group name, followed by several descriptive adjectives. Mark Catesby, for example, sought help from European savants for the name of the blue jay, which in his *Natural History* he called *Pica glandaria caerulea cristata* (magpie that eats acorns and is blue and has a crest). Scientific names were meant to be standard throughout the world, but such long names were obviously awkward in the extreme.

It was the Swedish naturalist Carl Linnaeus who, in the middle of the eighteenth century, developed a simpler system of naming. His *Systema Naturae* went through twelve editions in his lifetime. In it he tried to catalog all of nature, calling each species (beginning in the 1750s) by a double name, the group name (genus, or generic name) to be a capitalized noun, followed by the species name, usually an adjective agreeing in gender with the generic name. Thus what Catesby had called *Pica glandaria caerulea cristata* became simply *Corvus cristatus* (crested crow). (Linnaeus preferred to group it with the crows rather than the magpies; and the masculine word *Corvus* required a masculine adjectival ending, hence *cristatus*.)

Linnaean nomenclature came to be universally employed by scientists, though initially it met with little enthusiasm by non-scientists. One poet lamented that

> *. . . botanists of modern fame*
> *Newfangled titles chose to give*
> *To almost all the plants that live.*

Even today, many people are repelled by scientific names. Though few persons are familiar with Latin as such, many English words, and even more French, Italian, and Spanish ones, are based on Latin. Scientific names have no special rules of pronunciation; they can be pronounced as they would sound in one's own language. They are italicized not to frighten any-one, but simply because one usually italicizes words in another language in written text. It is easy enough to talk about the blackbirds in one's pasture, but an English visitor would be nonplussed—the blackbird of England is a thrush, not a black-bird in our sense. Their scientific names are different, and they are classified differently. Similarly, common mullein has dozens of names in Europe and America, but it is *Verbascum thapsus* everywhere. There are several million kinds of plants and ani-mals on earth. From a world point of view (the only view we moderns can afford to take) it is not useful to be imprecise about nature—which after all remains our major support sys-tem despite our vaunted technology.

There being only so many descriptive Greek and Latin nouns and adjectives, naturalists very early began to use the names of localities where the species was found (for example, *Neotoma mexicana* for the Mexican wood rat), or some aspect of their behavior (for example, *Canis latrans* for the coyote, *latrans* being Latin for "howling"). It also became common to name genera or species in honor of a person, often the discoverer, or simply someone who had made important contributions to that

area of science. A name based on that of a person is called an
eponym. Generic eponyms are created by adding a Latin ending
to a person's name, specific eponyms by adding the Latin pos-
sessives, *i, ii, anus,* or their feminine equivalents, *ae* or *ana.* I
have already cited several examples: genera such as *Bartramia,
Franklinia,* and *Kalmia;* species such as *michauxii, maximiliani,*
and *catesbeiana.*

To be properly established, a newly proposed name must be
accompanied by a description of the plant or animal, and specimens should be deposited in a reputable museum. Scientific
names are commonly followed by the name of the describer. For
example, it was Thomas Say who formally described the coyote
(though others had previously described it informally). The
name can thus be cited *Canis latrans* Say. In Peter Matthiessen's
words: "Until introduced to science by virtue of its description, a
species is not formally considered to exist, a technicality of small
consequence to the creature itself, which may have been getting
along quite nicely without a name for more than a million years."

Old-time naturalists recognized different kinds of plants and
animals on the basis of differences in color and form, and the
decision as to whether two entities differed enough to be considered separate species was sometimes arbitrary. Since description of a new species was considered a bit of immortality, some
naturalists worked very hard to find new species and attach
their names to them. Communication in the nineteenth century being nothing like it is today, naturalists were sometimes
unaware that someone else had already described a species or
genus, resulting in a duplication of names. There was also a tendency to apply names to what were no more than unusual variants or local populations. C. Hart Merriam, for many years chief
of the U.S. Biological Survey, described eight species of grizzly
bears!

Nowadays biologists recognize that members of a species
breed more or less freely among themselves but do not inter-

breed with members of other species (with rare exceptions). Sometimes localized populations of a species are sufficiently distinct to suggest that interbreeding with other races of the same species is limited; in this case these populations may be recognized by having a subspecies name attached to the species name. Much of the chaff resulting from the unnecessary proliferation of names has now been swept away (including the names of seven of Merriam's grizzly bears). Whenever there are several names that apply to the same genus or species, it is the first one to have been proposed that is normally used (the law of priority). Similarly, when two persons apply the same name to different organisms, it is the earliest application of the name that is accepted.

There has also been an effort to standardize the vernacular names of plants and animals. The U.S. Department of Agriculture has published a list of the common names of plants of economic value; the Entomological Society of America has published a list of insects of importance to humans; and the American Ornithologists' Union has standardized the names of American birds. Unfortunately such lists apply only within a given area. What we call the monarch butterfly, for example, is called the wanderer in Australia, though it is *Danaus plexippus* everywhere.

Names are always meaningful; they always bear a story. Eponyms bear two stories: that of the person who described the genus or species, and that of the person for whom it was named. Then there is the plant or animal itself—a third story. This is a book about eponyms and the stories they suggest. They often take us back to episodes and people we have forgotten, to our loss. Naturalists are individualists, preferring the wondrous products of organic evolution to the artificial world that humans have built at nature's expense. Sometimes they are cranky, a bit unworldly, even a bit mad. And there is always something to be said for the plants and animals that are their passion.

The nineteenth century was the time when Americans of European descent began to come to grips with the richness of the land they had wrested from its native inhabitants. It was a time when persons of scientific bent roamed the plains and forests and recorded their observations and discoveries. Virtually all left their names attached to plants or animals they encountered, and it is their stories that fill these pages. Now, as the twentieth century draws to a close, the scientific vision is inward, with electron microscopes and the sophisticated techniques of molecular biology; and outward, to space. Nature has become old-fashioned, some say moribund. Those of us who cherish what is left may find refreshment, in a jaded world, by revisiting those who reveled in discovery and by looking again at the plants and animals that excited them.

Abbot's Sphinx

*S*phinx moths are handsome and thoroughly admirable insects, usually seen at dusk dashing rapidly from flower to flower and uncoiling their long tongues to drink the nectar deep in the blossoms. "There is a high-bred tailor-made air about their clear-cut wings, their closely fitted scales, and their quiet but exquisite colors," wrote entomologist John Henry Comstock and his naturalist wife Anna Botsford Comstock. Often they are called hawkmoths because of their long, narrow wings and strong flight. At first glance they may be mistaken for hummingbirds; the larger ones are comparable in size to hummingbirds and their wingbeat is similarly so fast as to appear a blur. Some of the Native Americans of California believed that these moths turned into hummingbirds.

The larva of a sphinx moth is called a horntail, since the end of its body bears (in most cases) a hornlike projection. When the larvae rest, they rear the front of the body, facing forward rather majestically, suggesting to some imaginative people the face of the Egyptian sphinx. "But [wrote the Comstocks] we think they deserve the name independently of their habits because of the riddle they constantly propound to us as to why they wear this horn on the rear of the body instead of

Abbot's Sphinx

the head, where it ought to be in order to be of any use whatever as a horn."

A European species is called the "death's head sphinx"
because the markings on the back of the adult moth resemble a
skull and crossbones. Abbot's sphinx (*Sphecodina abbottii*) is by
no means as fearsome. It is a smallish species, as sphinx moths
go, having a wingspan of about two and a half inches. The front
wings are mottled brown, with jagged edges, admirably suited
for blending with the bark of trees on which the moths rest during daylight hours. When disturbed, they spread their wings
and display patches of yellow on their hind wings. This is
believed to be a "startle display," perhaps causing a predator to
shy away. In the evening, both males and females fuel themselves at the nectar of flowers, and following mating the females
fly to the foliage of grape or pepper vine, where the eggs are laid
and the larvae will feed.

Male Abbot's sphinx moths are usually seen flying at dusk,
while the females do not begin flying until well after dark.
Moths that fly at night cannot take advantage of the sun to
warm their bodies, although they must maintain a flight mus-

cle temperature of about one hundred degrees Fahrenheit in order to maintain rapid flight (about twenty-five to thirty wing-beats per second). Studies by Bernd Heinrich of the University of Vermont have shown that these moths are able to "warm their motors" sufficiently for flight by quivering their wings for a few minutes while at rest. Once in flight, they would be in danger of overheating if there were not also a means of keeping the flight muscles near the required temperature. They do this by increasing the flow of blood into the abdomen, which is more thinly scaled, permitting heat loss by convection. When the major blood vessel is artificially tied off, the moths overheat and can no longer fly. Most of this research was done not on Abbot's sphinx but on the larger tobacco hornworm moth, all too well known to tobacco and tomato growers.

John Abbot (1751–1840) was fortunate to have had such an attractive insect named for him (we shall forget that Abbot's bagworm, *Oiketicus abbotii*, also bears his name). Abbot has the unusual distinction of never having published anything in his life; yet he holds an important place in the annals of natural history. For more than a century after his death, almost nothing was known about him, although several speculative sketches of his life had been published. Then, in 1948, Charles Remington, later a professor at Yale University, discovered an autobiography of the first part of his life that had been locked in the archives of Harvard's Museum of Comparative Zoology.

Abbot told of his childhood in London, where his father was an attorney with an interest in art. "My peculiar liking for insects [he wrote] was long before I was acquainted with any method of catching or keeping them. I remember knocking down a Libelula [dragonfly] and pinning it, when I was told it would sting as bad as a rattlesnake bite."

Abbot's father arranged for young John to take drawing lessons. He became acquainted with Dru Drury, who had been president of the Linnaean Society of London. Drury allowed him to draw some of the insects in his large collection. For five years John Abbot had worked for his father as a clerk, but the work was "little to my liking when my thoughts was ingrossed by Natural-history." He found that he could sell specimens and paintings, and after studying specimens from distant parts of the world in Drury's collection, he began to "entertain thoughts of going abroad." A friend gave him a copy of Mark Catesby's *Natural History of Carolina, Florida, and the Bahama Islands*, and at the age of twenty-two he was off to America. He lived with a friend in Virginia for two years, where he prepared a collection of insects to ship back to London. But the ship was lost, along with his insects. He prepared another collection, but this was lost even before leaving port because of "a terrible September storm" that destroyed the ship.

"The times now becoming very troublesome [he wrote], & hearing that Georgia had not then joined the other Colinies, I joined [a friend] to come to Georgia together." He settled near Savannah and supported his wife and son by doing watercolors of insects and birds and selling his paintings to an agent in England who in turn sold them to wealthy English naturalists. Many of them came into the hands of Sir James Edward Smith, who in 1797 brought out a two-volume treatise titled *The Natural History of the Rarer Lepidopterous Insects of Georgia Collected from the Observations of John Abbot, with the Plants on Which They Feed*. Smith's book included many of Abbot's paintings of butterflies and moths in natural size and color, along with their caterpillars, pupae, and food plants, annotated with remarks supplied by Abbot, who never saw this book until long after it was published.

Sir James had earlier earned fame in England, and notoriety in Sweden, by making an offer to Linnaeus's heirs for his collections, "the greatest the world has ever seen," in Linnaeus's words. While others debated, Smith's offer of a thousand pounds was accepted, and the collections were soon on the brig

Appearance on their way to London. Included was a drinking cup made from the horn of a rhinoceros, ornately decorated and reputed to have magic powers. Perhaps it did, for Lady Smith lived to be 104.

Altogether Abbot made over five thousand watercolors, over a thousand of which were of birds. These he sold to individuals in America and abroad. The birds were posed rather stiffly, but the details were accurate and the species are easily recognizable even though some were not formally described until many years later. For example, Abbot figured what is clearly LeConte's sparrow, although the species was not named until Audubon did so in 1844. Ornithologist Walter Faxon wrote that if Abbot "had secured the speedy publication of this remarkable collection of drawings, with a suitable accompaniment of text, his name would be famous in the annals of American ornithology." Indeed, Abbot never published anything, and had Smith not made known his paintings of Lepidoptera, his name might have faded into obscurity.

Abbot also supplied specimens to collectors in the United States and abroad. John L. LeConte purchased beetles from him, paying him six dollars per hundred specimens. Alexander Wilson paid him forty-five dollars a specimen for birds and made frequent reference to Abbot in his *American Ornithology*. (LeConte and Wilson will both appear in later chapters.) Among the birds described by Wilson on the basis of specimens sent by Abbot were the field sparrow, the solitary vireo, and the black-billed cuckoo. Wilson also used information supplied by Abbot in his write-ups of the ruby-throated hummingbird, the painted bunting, and several other species.

On a trip through the South, Wilson called on Abbot at his home near the Altamaha River. It must have been a remarkable meeting between the somber Wilson and the older but still youthfully lighthearted Abbot. They doubtless had much to talk about, but what was actually said we can only guess. After

Wilson's untimely death, Abbot sent field notes to George Ord, who was completing Wilson's *American Ornithology*. Abbot wrote to Ord that he had "retired into the Country . . . in hopes the mad and destructive Ambition of the rulers of the world can but little interfere." His ties to England had been interrupted by the War of 1812.

Many of Abbot's paintings eventually found their way to the British Museum and are still housed in the zoology library of the museum. Others, purchased by wealthy Americans, found their way to the Houghton Library of Harvard University, to the Smithsonian Institution, and to the University of Georgia. Some of the bird specimens were acquired by William Bullock for his London museum, the contents of which were auctioned off in 1819 (more about Bullock in the next chapter). Many came into the hands of Lord Edward Stanley, an ardent amateur zoologist, and some of the specimens are still preserved in the Merseyside County Museum of Liverpool.

In an essay of appreciation, Robert Dow described Abbot as "an untutored optimist, with a constitutional smile, who looked forward only to the day's reward, who had talent with the brush, who had the assiduity to rear every insect species he could for fifty years." When Abbot lost his hearing and much of his sight in old age, he hired local boys to collect specimens for him. Both his wife and his son had predeceased him by many years.

Among others who acquired material from Abbot, either directly or indirectly, was Englishman William Swainson (1789–1855), who was also primarily an illustrator. Between 1820 and 1823 he produced three volumes titled simply *Zoological Illustrations*, which included 182 plates, all drawn and hand colored by himself. It was here that he figured and named Abbot's sphinx, on one of his best executed plates. Like Abbot, Swainson had from his youth liked to do nothing but paint natural objects; later he developed a new and advanced technique of lithography.

As a young man, Swainson spent several years in Sicily, where he spent several weeks collecting specimens with Constantine Rafinesque. In 1816 he tried to find support for an expedition to Brazil, but failing in that he set off on his own. For nearly two years he traveled in the American tropics, mostly in eastern Brazil, which was still little explored for its natural history. (It was not until thirty years later that W. H. Edwards made his much better known trip to Brazil, followed shortly by the even better known trip of Henry Walter Bates and Alfred Russell Wallace.) Swainson returned to England with 760 bird skins, including many hummingbirds, as well as 20,000 insects and many drawings of Brazilian wildlife.

Swainson became a good friend of William Leach (1790–1836), assistant keeper at the British Museum of Natural History. When Leach assumed the post, he was appalled by the neglect of the collections, so he and Swainson dragged many of the more hopeless specimens out of doors and had several foul-smelling bonfires. Leach lived in two rooms at the museum. For exercise he ran up and down the stairs and now and then leaped over the back of a stuffed zebra "with the lightness of a harlequin" (Swainson's words).

Leach worked very hard to improve the collections and exhibits, often late into the night, and over time he became increasingly pale and emaciated. His research (primarily on crustaceans and mollusks) bore the marks of his eccentricities. Among other things, he became preoccupied with a person named Caroline and based nine generic names on her name or anagrams of it. (Was it Queen Caroline, or a mysterious paramour?) When he was forced to resign in 1822 after a complete mental breakdown, Swainson applied for his position, but he was passed over for a less qualified person who had better connections.

In 1819 Leach had bought a mounted petrel at the auction of specimens from Bullock's Museum in London. The price was five pounds, fifteen shillings—rather a lot for a black bird only eight inches long. The following year a visitor to the British Museum decided it was a previously unknown species and described it as *Procellaria leachii*. It turned out that French naturalist Louis Vieillot had described it from a different specimen three years earlier, and his name therefore had priority. However, the common name Leach's storm-petrel is still the one in use. Although first discovered in Europe, these birds are sometimes seen along our coasts, flying close over the water and snatching fish from the waves.

After being turned down by the British Museum, Swainson married and began to support his family through his writings and illustrations. He never visited North America, but he was much involved with its fauna when he collaborated with John Richardson on the volume on birds in the *Fauna Boreali-Americana; or the Zoology of the Northern Parts of British America* (1831).

Swainson became acquainted with John James Audubon when the latter first visited London in 1828. The two got on well at first, and for a time used the same engraver for their plates, R. Havell & Son. They went to Paris together and visited the Jardin des Plantes to meet the great zoologist Georges Cuvier, who introduced them at a meeting of the French Academy of Sciences. For a time Swainson and Audubon talked of working together, but Swainson later quarreled with Audubon over access to bird skins and to data needed for his publications.

It was not unusual for Swainson to quarrel with his contemporaries, many of whom held him in high regard as an illustrator but did not rate him as highly as a naturalist as he felt he deserved. "Modesty [wrote a biographer] was never Swainson's problem." In his 1840 book *Taxidermy, with the Biographies of Zoologists*, he presented sketches of the lives of naturalists, living and dead. He gave himself much the longest entry (fourteen pages) and included his own portrait as a frontispiece (Audubon received one page and was cited simply as an illustrator).

Swainson did not feel he was appreciated in England, and after the death of his wife he sold his collections and moved to New Zealand with a new wife and his five children. But the Maoris were revolting against their British oppressors, and it was impossible for him to venture far from his home. Cut off from the world of nature, he must have found little solace for his declining years. Swainson's thrush, Swainson's hawk, and Swainson's warbler are all named for him, a man who, like Abbot, had no formal training in natural history or medicine but had a passion for drawing and painting, very effectively, a variety of natural objects. Both will be well remembered for the birds and insects named for them.

Anna's Hummingbird

Anna's hummingbird was not named for or by an American. Rather it was named by French naturalist René Primevère Lesson (1794–1849) for Anna, Duchess of Rivoli and wife of Prince Victor Massena, an amateur French ornithologist. Audubon met her in Paris in 1828 and described her as a "beautiful young woman . . . extremely graceful and polite."

Anna's hummingbird is unique in that it migrates little if at all, spending the entire year in a narrow strip along the Pacific coast. Its range includes several universities, and not surprisingly it has become a favorite of researchers inquiring into the energy requirements of these diminutive bundles of intensity. They have found that an Anna's hummingbird's rate of oxygen consumption (as a measure of body metabolism) exceeds that of any other animal measured under similar conditions; but while in flight its oxygen consumption is five times that while perched. To obtain the necessary calories, an Anna's hummingbird must visit about a thousand fuchsia blossoms a day for nectar. But this is balanced by the fact that hummingbirds become more or less torpid at night; their oxygen consumption is very low, and their temperature may drop as low as seventy degrees Fahrenheit, as compared to a normal temperature of slightly over one hundred. All of this reinforces what we already knew: that hummingbirds are quite the most remarkable of birds.

Early visitors to the New World were astonished by them (they do not occur in the Eastern Hemisphere). Here is Spanish explorer-naturalist Gonzalo Fernández de Oviedo, writing in 1526: "There are found . . . certaine birds, so little that the whole bodie of one of them is no bigger than the top of the biggest finger of a man's hand. . . . This Bird, besides her littleness, is of such velocitie and swiftness in flying, that who so seeth her flying in the aire, cannot see her flap or beate her wings. . . . Their Feathers are of manie faire colours, golden, yellow, and greene."

Oviedo was writing from the tropics, but in the United States there are over a dozen resident species, as well as several others that occur casually. To observe any one is to recapture some of Oviedo's astonishment.

Abert's Squirrel

Surely a word must be said for Abert's squirrel, one of the most elegant of its tribe. Otherwise known as tassel-eared squirrels, these animals occupy a limited range along the Rocky Mountains, from extreme southern Wyoming to central New Mexico and Arizona. In the winter they grow "earmuffs," brushes of long hairs that cover their ears and can often be seen being deflected by the wind when the squirrels are cavorting in the trees. I am lucky to live within the range of these attractive squirrels and can report that on cold mornings the ear tassels are sometimes tipped with hoarfrost, so far do they extend beyond the warmth of the ear proper. Abert's squirrels feed mostly on the seeds of ponderosa pines but, like most squirrels, will not pass up a meal from a bird feeder. Most of the squirrels where I live are solid black, not exactly a protection against the local hawks when a squirrel perches on top of a snow-covered boulder. Farther south (and occasionally here, at the northern edge of the range) the fur is gray above and white beneath.

Abert's squirrel was named by Dr. Samuel Woodhouse, about whom I'll have more to say in a later chapter. Woodhouse had been chosen by Colonel John James Abert, chief of the U.S. Topographical Engineers, to serve on several expeditions to the West in the late 1800s. Though an army man, Abert was a keen observer of nature. He once entertained Audubon at his home in Washington, D.C., and was able to assure Audubon that rattlesnakes do sometimes climb trees, as Audubon had shown in one of his paintings (for which he had been castigated by some of his rivals). When Woodhouse returned to Philadelphia, he described several of the animals he had collected in the West and chose to name the squirrel *Sciurus aberti* after his superior officer. One wonders if Colonel Abert appreciated how special an animal bears his name.

Agassiz's Turtles

Louis Agassiz (1807–1873), a Swiss by birth, arrived in Boston in 1846 and began to conquer the country with his magnetism and his eloquent espousal of natural science. Already he had authored a monumental treatise on fossil fishes as well as major publications on living fishes. His belief that parts of the Northern Hemisphere had been covered by successive periods of glaciation was controversial, but time has vindicated him.

Established at Harvard, Agassiz began to solicit fishes, turtles, and other zoological specimens from all over the world, and within a few years he had founded his Museum of Comparative Zoology to house all this material and to serve as a research and teaching museum. He was an imposing figure on the lecture platform, with a head rather too large for his body, made seemingly still larger by a receding hairline. There was always a blackboard at hand. Like no one else he had the ability to make even the most recondite zoological details clear to students and to ordinary citizens alike. So popular were his lectures that Ralph Waldo Emerson complained that "something should be done to check the rush towards natural history."

Agassiz planned a series of monographs on American natural history, each so well illustrated as to appeal also to nonscientists. Over twenty-five hundred people subscribed to the first volume, even though it dealt with those homely and sluggish animals, the turtles. Only three more of the projected ten volumes were ever completed, as Agassiz soon found that his duties as a teacher, administrator, and public figure consumed most of his time. He moved in the best of circles; he was a friend of the Lowells and married a cousin of the Cabots. Henry Wadsworth Longfellow and Oliver Wendell Holmes became good friends. Henry David Thoreau collected turtle eggs for him in Concord and discussed the copulation of turtles with him over dinner at the Emersons. "A man on a heroic scale," wrote William James, who as a young man traveled with Agassiz to the Amazon.

On Agassiz's fiftieth birthday, Longfellow presented him with a poem of praise, and upon his departure for Brazil, Holmes composed "A Farewell to Agassiz":

> *May he find with his apostles*
> *That the land is full of fossils,*
> *That the waters swim with fishes*
> *Shaped according to his wishes,*

> *That every pool is fertile*
> *With fancy kinds of turtle. . . .*

Agassiz often wrote for the *Atlantic Monthly*, and when he died James Russell Lowell composed a poem of tribute that filled eleven pages of that journal. Rarely have the arts and sciences concorded so fully as in the career of Louis Agassiz.

But to return to his turtles. In his monograph he described several species, including the yellow mud turtle and the Pacific hawksbill. In time several came to be named for him, including Agassiz's soft-shelled turtle and the desert tortoise (*Gopherus agassizii*). The latter is much in the news these days, as it is classified as a threatened species. Its habitats are being lost to sun-belt homes, and individual turtles are being used for target practice or crushed by off-road vehicles as they roar across the desert in clouds of dust. Desert tortoises do not reach sexual maturity until they are fifteen to twenty years old and they may live another forty years. Like most turtles and tortoises, their lives are slow-paced, and they require nothing of their environment but a few blades of grass. Researchers have placed miniature radio transmitters on the backs of some individuals and discovered that they have a home range of forty to fifty acres. These unlovely creatures are admired by Californians to the extent that they have been designated state reptile, and the Desert Tortoise Council has been formed to assist in their protection. John Alcock devoted a chapter to them in his delightful book *Sonoran Desert Spring:*

> To see a tortoise with wrinkled neck and solemn eyes, moving like an animated rock, is an essential part of the experience of the desert. The removal of even a single adult extinguishes a presence that was meant to persist for years to come and snuffs out a prehistoric spark of life in a spartan environment where life, so hard-won, should be celebrated.

The desert tortoise was described and named for Agassiz by James Graham Cooper (1830–1902), a physician who was attached for a time to the Pacific Railroad Survey and to the Geological Survey of California. He was an ardent naturalist and published reports on mammals, plants, and shells, as well as a pioneering *Ornithology of California*. The Cooper Ornithological Club was named for him, but Cooper's hawk was named for his father, William Cooper, one of the founders of New York's Lyceum of Natural History, now called the Museum of Natural History.

Baird's Sandpiper

There are few if any roles in nature that are not filled by actors specially fitted to fill them. Sandy or muddy margins of ponds and lakes and sea beaches are the habitats of small crustaceans, fly larvae, beetles, and other small organisms, and a whole clan of birds has evolved to exploit them as a source of food: the sandpipers. With their long bills and broad feet they run over the sand, constantly probing, now and then taking flight in a swirling cloud of beating wings. "Peeps" they are often called, since "peep" is about all they have to say during their migrations along our shores or inland waterways. Most of us see them only during their spring and fall migrations, as the majority of the species breed in the Far North.

Sandpipers are the bane of bird-watchers; to a casual observer the dozen or more species all look and act very much alike. Baird's sandpiper (*Calidris bairdii*) is rather a generic sort, larger than some, smaller than others, with the gray-buff plumage common to many species. Perhaps Spencer Fullerton Baird (1823–1887), a dominant figure in American natural history for many years, deserved to have a more spectacular bird named for him (Baird's sparrow is also a fairly nondescript "little brown bird"). But he did have an Alaska glacier and inlet named for him, as well as a whale and many other mammals.

Of course there is something to be said for every living thing, and Baird's sandpiper is no exception. As migrants they rival the arctic tern, spending the summer in extreme northern Canada and Alaska and the northern winter in Chile and Argentina, all the way to Tierra del Fuego. During migration, they occur along ponds and reservoirs in the plains, even occasionally in the mountains. Roger Tory Peterson found them high in the Andes, up to sixteen thousand feet in elevation.

William Drury of the Massachusetts Audubon Society studied the sandpipers on their breeding grounds on Bylot Island, just north of Baffin Island and well north of the Arctic Circle. The birds arrive in early June and occupy places on the tundra where the snow has melted back, generally on dry slopes overlooking the coastline or areas of fresh water. Plenty of small insects are available here as food for themselves and their chicks.

Baird's Sandpiper

Once on the nesting area, the males establish territories which they advertise with displays that are surprisingly elaborate for birds of such modest plumage. They fly up in the air thirty to fifty feet and produce a "froglike trilled song" that is continued as they circle and glide back to the earth over the next few minutes. Dr. Drury distinguished three kinds of flight patterns: a slow wingbeat, a "butterfly display" (wings held at a high angle to the body), and a "moth display" (with a shallow, quivering wingbeat). When they land, the males produce a "ground song," holding one wing high over their body. Before mating, the male and female crouch facing one another "like game cocks," and during mating the male pulls feathers from his mate's head.

Nests are built on the ground of pebbles, mosses, and lichens, and four brownish eggs with darker spots are laid. After three weeks the chicks appear and are able to feed themselves almost immediately. But they remain near the nest for a few days and the adults protect them by performing diversionary tactics similar to those of many sandpipers and plovers.

Perhaps, after all, the way these sandpipers hold their own *is* a tribute to so notable a person as Spencer Baird, assistant secretary of the Smithsonian Institution for twenty-eight years and for several years its secretary (that is, director). As a youth in Carlisle, Pennsylvania, Baird corresponded with Audubon, who encouraged him in his studies. Audubon was then working on his volume on quadrupeds, and he asked Baird "please to collect all the Shrews, Mice, (field or wood), rats, bats, Squirrels, etc., and put them in a jar in common Rum, not whiskey, brandy or alcohol. All of the latter spirits are sure to injure the subjects."

Later Baird and Audubon became well acquainted, and Baird became "hooked" on birds, even though his older brother William assured him that "no means of livelihood . . . is to be obtained in America from ornithology." (How many young naturalists have heard these words!) When the Wilkes Expedition returned from the South Seas in 1842, Baird was anxious to examine its findings, then housed in the U.S. Patent Office. He visited his brother, who was with the Treasury Department in Washington, and became acquainted with the expedition's geologist, James Dwight Dana, and other naturalists. Dana was one of several influential people (including Audubon) who supported Baird's application as assistant secretary of the Smithsonian soon after it was founded in 1846.

Before that, Baird taught for several years at Dickinson College, where he often took his students on field trips to study and collect specimens. In 1846 he married Mary Churchill, daughter of the inspector general of the army. In a letter to Dana he extolled the virtues of his wife, surely the perfect wife for a naturalist.

> Not the least fear of snakes, salamanders, and such other Zoological interestings; cats only are to her an aversion. Well educated and acquainted with several tongues, she usually reads over all my letters, crossing i's and dotting t's, sticking in here a period, and there a comma. . . . In my absence, she answers letters of correspondents, and in my presence reads them. She transcribes my illegible Mss., correcting it withal, and does not grudge the money I spend on books. In addition to all these literary accomplishments, she regulates her family well (myself included) and her daughter is the cleanest and most neatly dressed child in town.

Their daughter was named Lucy, after Audubon's wife. Lucy's warbler (*Vermivora luciae*) was later named for her by one of Baird's protégés, James Graham Cooper, whom we met as the describer of the desert tortoise.

Baird's initial salary at the Smithsonian was $1,500 a year, small for a Washington resident even in 1850. He worked ten to fifteen hours a day and produced important volumes on birds while training and supporting several young naturalists who traveled widely and added greatly to the Smithsonian's collections. His thirty-seven years at the Smithsonian embraced the years of the final opening of the West, when railroads were built; boundaries, rivers, and mountains were surveyed; and homesteaders tried their luck on virgin lands. Government-sponsored expeditions included those of Clarence King (1867–1873), F. V. Hayden (1868–1871), John Wesley Powell (1869–1874), and George Wheeler (1871–1874). Baird saw to it that whenever possible naturalists were attached to each expedition and that the specimens they collected would make their way to the Smithsonian. He corresponded with virtually all the naturalists of his day, and many of his letters are still preserved in the Smithsonian's archives. Although he did not travel extensively himself (partly because of a nagging heart problem), he was a central figure in natural history investigations of the latter half of the nineteenth century, and his name will appear many times in these pages.

Baird was a modest and generous person, but his relationship with Louis Agassiz, a dominant figure in American science at the time, was not always the best. Agassiz was forever borrowing specimens from the Smithsonian and forgetting to return them, and Baird became reluctant to part with more specimens. When Agassiz's protégé Charles Girard fled to Baird in 1850, Agassiz was unforgiving. He opposed Baird's appointment to the National Academy of Sciences in 1864, but he failed to block the nomination.

For years, Baird argued that a national museum be founded to contain the Smithsonian's natural history holdings and to serve as a research center for studies of biological diversity. But it was not until 1875 that Congress recognized the need, and not until 1881 that the museum building was completed. The National Museum of Natural History has grown steadily through the years not only in its holdings but in its stature as one of the great research museums of the world, a tribute to Spencer Baird's energy and dedication.

Fish had always been a major interest of Baird's, and for a time he and Agassiz had considered coauthoring a monograph on the natural history of North American fishes. Baird and his family spent many summers on the coast, partly for their health. In 1871 he wrote to the congressional committee on appropriations, pointing out "the great diminution in the numbers of the fish which furnish the summer food supply of the coast. . . . It is necessary that a careful, scientific research be entered upon, for the purpose of determining what should really be done; since any action presupposes a knowledge of the history and habits of the fish of our coast, that, I am sorry to say, we do not at present possess."

As a result of Baird's influence, Congress shortly appropriated funds for the establishment of the U.S. Fish Commission, with Baird as commissioner (at no additional salary). Baird had been visiting Wood's Hole, Massachusetts, since 1863 and had built a summer home there, so it was logical that the commission's headquarters be located there (in fact, for a time it occupied part of Baird's house). Baird remained commissioner until his death in 1887. He had hoped to establish an institute of biology at Wood's Hole, but Congress was unwilling to supply the funds. However, with financial help from several wealthy friends, Baird laid the groundwork for the founding of the Marine Biological Laboratory, which was formally dedicated in 1888, the year after his death.

It was a protégé of Baird's, Elliott Coues (1842–1899), who named Baird's sandpiper on the basis of specimens collected at Great Slave Lake, in northern Canada, by Robert Kennicott. Coues dedicated the species to Baird "as a slight testimonial of respect for scientific acquirements of the highest order, and in grateful remembrance of the unvarying kindness which has ren-

dered my almost daily intercourse a source of great pleasure."
This appeared in Coues's first publication, a monograph on
North American sandpipers; he was only nineteen when it was
published.

Like Baird, Coues (pronounced "cows") was trained in
medicine, but he had been a bird enthusiast from childhood.
While he was still in college, he traveled to Labrador to collect
birds for Baird. When he returned, he published "Notes on the
Ornithology of Labrador," filled with vivid accounts of the birds
he had observed. His description of his first encounter with the
Atlantic puffin reveals something of his youthful enthusiasm—
and something of the ruthlessness of many early bird collectors.

> Hardly had our boat touched the shore when we leaped out,
> guns in hand, and at once scattered over the island. As we
> advanced along the sides, the affrighted birds darted past us
> like arrows, issuing from their burrows beneath our feet and
> around us, and making directly for the water. . . . I took my
> stand on a flat rock, and in less than an hour a pile of Puffins,
> more than I could carry, lay at my feet.

Later, as an army surgeon, Coues was assigned, with Baird's assistance, to posts in areas where the birds and mammals were poorly
known. While in Arizona he wrote to Baird that "the Apaches are
so hostile and daring that considerable caution will have to tinge
my collecting enthusiasm, if I want to save my scalp." In Arizona
he collected a new species of warbler and asked Baird to name
it after his sister Grace. Grace's warbler (*Dendroica graciae*) is
an attractive inhabitant of coniferous forests of the Southwest.
One of the officers stationed with Coues remembered that
during a march to a new post in Arizona, Coues,

> clad in a corduroy suit of many pockets and having numerous
> sacks and pouches attached to his saddle . . . regularly rode

out of column every morning astride of his buckskin-colored mule, which he had named Jenny Lind on account of her musical bray. Rarely did we see him again until we had been some hours in the following camp but we sometimes heard the discharge of his double-barreled shotgun far off the line of march. He usually brought in all his pockets and pouches filled with the trophies of his search, and when he sat upon the ground and proceeded to skin, stuff and label his specimens, he was never without an interested group of officers and men about him.

After a year and a half, Coues felt the strain of being so far from libraries and museums and was able to transfer to posts in the Carolinas and in Maryland. With ready access to the Smithsonian, he began his *Key to North American Birds*, which was destined to be his most influential publication. It went through five editions in his lifetime and one after his death. Coues illustrated his key with many sketches of his own making, useful but of variable quality. A few years later he discovered a young Cornell student who, despite his father's objections, wanted to spend his life painting birds. His name was Louis Agassiz Fuertes, and he went on to surpass Audubon in the beauty and liveliness of his paintings. Fuertes illustrated the last two editions of the *Key*.

In 1872 Coues left for "another period of exile" in the West, this time in the Dakotas and Montana, where he worked with the Survey of the 49th Parallel. Here he collected not only birds but mammals, fish, and insects. Entomologists W. H. Edwards and P. R. Uhler, who will appear in later chapters, described many new insects from Coues's collecting. A few years later Coues was appointed secretary and naturalist to the U.S. Geological Survey of the Territories, under F. V. Hayden, an assignment that allowed him to spend much of his time in Washington. His *Birds of the Northwest* (1874) and *Birds of the Colorado Valley* (1878) contain some of the best life-history studies of birds ever published. Coues believed that "it is possible to make natural history entertaining and attractive as well as instructive," and his writings bear out his belief.

In 1881 Coues retired from the army and accepted a professorship at Columbian (now George Washington) University. He continued to publish prolifically on birds and mammals and on philosophy and spiritualism. The titles of some of his publications are intriguing: *The Harmony of Scientific Knowledge and Religious Faith* and *Can Matter Think?* Coues had always been interested in the history of scientific discovery, and late in life he edited a four-volume account of the journals of Lewis and Clark and a three-volume account of the expeditions of Zebulon Pike. He contracted his final illness while traveling through the Southwest, following the route of Francisco Garcés, a Franciscan padre who came to the mission of San Xavier, in southern Arizona, in 1767. Garcés visited Native American tribes as far away as the Havasupai, in Grand Canyon, and into California. Although he spread goodwill wherever he went, in 1781 he was clubbed to death during an uprising of the Yumas. Garcés often made notes on natural history, and the publication of an edited version of his diaries was one of Coues's last acts.

Coues was one of the founders of the American Ornithologists' Union and helped to prepare the first two editions of their *Check-List of North American Birds*. He was a strong believer in recognizing local races of birds and mammals by the use of subspecies names—a controversial practice at the time although now well established. Coues was well aware of his own worth and often highly critical of others. From time to time he feuded with naturalists as eminent as Robert Ridgway and with clerics as important as Henry Ward Beecher. He was married three times and was reputed to have had "indiscretions" from time to time. Altogether he was a colorful and controversial figure, with a full beard and strong opinions. Coues described six genera of birds and four genera of mammals, as well as several species and subspecies. He himself had several named for him, including Coues's flycatcher and Coues's rice rat. It is tempting to believe that the rat was named for him by one of his enemies, but that is not the case.

Thomas Mayo Brewer and the Sparrow War

The Sparrow War of 1874–1878 has never made it into history books, but it engendered a good deal of passion in ornithological circles. Elliott Coues was once again feuding, this time with Thomas Mayo Brewer (1814–1880), a Harvard-educated physician who preferred birds to doctoring. The house sparrow had been introduced from Europe in 1852 by persons who liked to be reminded of the Old Country and who thought the sparrows would clean up the cankerworms in city trees. But of course the sparrows are seed-eaters and scavengers and feed mainly on the ground. In Coues's words, they had soon "overrun the whole country, and provide a nuisance without redeeming quality." Brewer was a defender of the sparrow, and he had strong backing from the eloquent clergyman Henry Ward Beecher. For several years newspapers and scientific journals were filled with blasts and counterblasts. Brewer wrote in the *Washington Gazette* that Coues's statements regarding the sparrow were an example of "a lie well stuck to being as good as the truth." After Brewer's death, Coues wrote of Brewer's accomplishments but said that he had "made a fool of himself about the sparrow for years." The sparrow won the war; Coues's suggestions that the sparrows be controlled were never implemented to any degree, even though most naturalists as well as many householders agreed that the sparrows were a bit of a pest.

Brewer was a leading ornithologist of the generation just before Coues. The publication in 1857 of his book *North American Oölogy*, filled as it was with colored plates of bird's eggs and with notes on life histories and distributions, established him as the country's leading oölogist. In 1875 he collaborated with Spencer Baird and Robert Ridgway in a monumental three-volume compendium, *The History of North American Birds*. During Audubon's later years, he and Brewer were good friends, and Audubon made much use of information provided by the younger man. Audubon honored Brewer by naming two birds for him. Brewer's blackbird is a sleek if somewhat noisy denizen of farms and fields of the West. Brewer's duck turned out to be a hybrid between a mallard and a gadwall, as Audubon half suspected, so the name is no longer used. John Cassin (whom we'll meet in the next chapter) named Brewer's sparrow, an inhabitant of western prairies and sagebrush flats with a song of unusual exuberance for a relatively nondescript bird. Spencer Baird named the beach vole *Microtus breweri*, and Audubon's friend and collaborator John Bachman named the hairy-tailed mole *Parascalops breweri*. So Brewer is well remembered despite his devotion to the pesky house sparrow.

Bell's Vireo

Audubon named birds after many of his friends; he had so many species to name that doubtless he felt fortunate that he had so many friends. Bell's vireo (*Vireo bellii*) he named for John Graham Bell (1812–1889), a New York taxidermist who accompanied him on his journey up the Missouri River in 1843. Well known only to ardent bird-watchers, Bell's vireo is declining in parts of its range as a result of parasitism by cowbirds. A few years after his trip with Audubon, Bell collected birds in

California, and John Cassin (1813–1869) named one of them *Amphispiza belli*, commonly known as the sage sparrow.

Bewick's Wren

Audubon named Bewick's wren for Thomas Bewick (1753–1828), author of A *History of British Birds*. Audubon visited Bewick and "found him at all times a most agreeable, kind, and benevolent friend." Audubon was grateful that Bewick found him several new subscribers to his *Birds of America*. Bewick was a master of wood engraving who was versatile enough also to provide illustrations for *Aesop's Fables* and Robert Burns's *Poetic Works*. His book on birds was very popular, though it added little that was new. Still, it is good that he is remembered in the name of a sprightly wren of western woodlands and streamsides.

Barrow's Goldeneye

Barrow's goldeneye is a very different bird with a very different history. Diving ducks of exquisite coloration, these birds breed on inland lakes of the Far North, rarely as far south as Oregon, and spend the winters on both coasts. Sir John Barrow (1764–1848), secretary of the British admiralty, was a founder of the Royal Geographic Society and author of A *History of Arctic Voyages*. He directed several expeditions from his offices in London, including those of James Clark Ross, Edward Sabine, and John Franklin (all of whom will appear on later pages). Barrow Strait, Point Barrow, and the village of Barrow, Alaska, are all named for him. Barrow's goldeneye was described by William

Swainson on the basis of a specimen taken outside its usual range, in the Rocky Mountains, by Thomas Drummond (we'll meet him, too, in a later chapter).

Bullock's Oriole

Bullock's oriole is now considered only a subspecies of the northern oriole, but it would be unfortunate to pass up such a startling, halloween-colored bird and so curious a figure as William Bullock (1775–?1849). Bullock was an important figure in England at the beginning of the nineteenth century, describing himself as a "Silversmith, Jeweller, Toyman, and Statue Figure Manufacturer." He also collected natural history items, and in his museum a great miscellany of objects was displayed. The museum was started in Sheffield, but proved so successful it was moved to Liverpool and then to London's Piccadilly Square. In a publication describing his museum, Bullock acknowledged material from "Mr. Abbot, the naturalist, of Savannah in America." When Sir Joseph Banks returned from his voyage with Captain Cook, he permitted most of his zoological and ethnological materials to be dispersed widely, and some of them came to Bullock's Museum. In a way, the museum was a forerunner of P. T. Barnum's Museum in New York, and it is interesting that three decades later Barnum himself rented one of Bullock's buildings to display Tom Thumb to Londoners.

Bullock's standards for mounting specimens were high, and he published A *Concise and Easy Method of Preserving Objects of Natural History.* In 1812 he took part in a wild chase by boat after one of the last great auks of the British Isles. He failed to get it but later acquired one for his museum. In 1819 he suddenly decided to sell his collections at auction; they then comprised thirty-two thousand "curiosities" and over three thousand stuffed birds. The sale took twenty-five days, with Bullock himself serving as auctioneer. Bidders included William Swainson, William Leach, Sir Walter Scott, and others from Germany, Holland, France, and Austria. Bullock was then off to Mexico, where he speculated in mines and incidentally collected several previously unknown birds. On his return to England he published an account of his adventures, *Travels in Mexico*. He later went to Central and South America. But with that he disappeared completely, and no one knows where or how he died or the date of his death. It was William Swainson who preserved his name in that of one of our most brilliantly colored birds.

Bartonia

In the eighteenth century and before, the serious study of plants was primarily a search for new medicines and palliatives, and the common names of many plants reflect this: fleabane, lousewort, pleurisy root, alumroot, and so forth. Not surprisingly, America's first native-born botanist was much concerned with this subject. Benjamin Smith Barton (1766–1815) was born in Pennsylvania but trained in medicine in Europe. He settled in Philadelphia, where he became professor of materia medica at the University of Pennsylvania and one of the city's more prominent citizens. His *Elements of Botany* (1803) was a somewhat rambling, philosophical book, but it was the first textbook on the subject in the country. At times Barton employed Frederick Pursh and Thomas Nuttall to collect plants for him (both went on to notable careers as botanists, and will appear in later chapters). Thomas Jefferson and William Bartram were good friends of Barton's. When Jefferson wanted to

give Meriwether Lewis a "crash course" in natural history prior
to his expedition to the West, it was to Barton that he sent him.
Lewis carried *Elements of Botany* all the way to the Pacific and
back. Barton himself was frail and unable to travel much; he
died of tuberculosis when he was forty-nine.

In 1812 his former assistant Frederick Pursh dedicated a
spectacular genus of plants to him: *Bartonia*. These are called
evening stars; the large, sharply pointed petals spread open in
the evening or on cloudy days like many-pointed stars. There
are several species, some locally common. These showy and
unusual plants seem a fitting tribute to a person so important
in the history of botany. Pursh went on to publish a North
American *Flora* in 1814, and this widely used book established
Bartonia as both the common and the scientific name of the
plants.

Unfortunately the name *Bartonia* had been applied eleven
years earlier to a much less impressive plant by Barton's friend
Henry Muhlenberg. Since nowadays botanists insist on strict
priority, the name *Bartonia* is now properly applied to a little-
known genus of the gentian family having small white or yellow
flowers. Presumably Pursh was unfamiliar with Muhlenberg's
earlier use of the name. Barton had always held a good opinion
of himself, and doubtless would have preferred Pursh's *Bartonia*
to Muhlenberg's.

Britton's Skullcap

A century later, Nathaniel Lord Britton (1859–1934) authored
several major books on botany, including *An Illustrated Flora of
the Northern United States, Canada, and the British Possessions*
(coauthored with Addison Brown). Britton was director of the
New York Botanical Garden, Brown its president. The three-

volume work, which has been updated several times, includes excellent sketches of every plant within the region covered, a blessing to those of us who admire plants but are not trained in botany. Mount Britton in Puerto Rico was named for Britton in recognition of his book *Botany of Puerto Rico and the Virgin Islands*. Many plants bear his name, including Britton's spike-rush, *Eleocharis brittonii,* and Britton's bush-clover, *Lespedeza brittonii*. Britton's skullcap (*Scutellaria brittonii*) springs up in unexpected places around my home each summer. It grows only a few inches tall, but the bright blue flowers are most welcome. The name "skullcap" evidently refers to the uppermost petal, which perches like a cap over a blossom remotely resembling a skull. These are members of the mint family, but they seem to have none of the minty odor of most members of that family.

Clarkia

From mid-May to mid-June 1806, the Lewis and Clark Expedition, returning from the Pacific, remained stalled on the Clearwater River, waiting for the snows to melt in the Bitterroot Mountains (of present-day Idaho). They were living on roots and on horse and dog meat; many in the party were ill, and all were depressed and homesick. Most of the time was devoted to gathering enough food for their immediate needs as well as enough to sustain them when they could finally cross the mountains. But good naturalists that they were, Meriwether Lewis and William Clark took the occasion to collect specimens and record observations on the life around them. Among the birds seen were those now known as Clark's nutcracker, Lewis's woodpecker, and western tanager. Lewis studied the carcasses of several bears the hunters brought in and easily distinguished between those of black bears and those of grizzlies.

Among the plants they collected and eventually sent to Philadelphia were camas lilies, mariposa lilies, and a remarkable relative of the evening primrose that later came to be named *Clarkia*. Lewis wrote in his journal: "I met with a singular plant today in blume of which I preserved a specemine; it grows on the steep sides of the fertile hills near this place. . . . [The blossom] consists of four pale perple petals which are tripartite, the central lobe the largest and all terminate obtusely. . . ."

Nowadays this species of *Clarkia* is often called ragged robin, since the blossoms appear to have been torn into shreds. It is perhaps the most striking and unusual of any flowering plant. These are annuals that grow about a foot tall, mainly in showy patches among sagebrush and bunchgrass. In 1814 Frederick Pursh named the species *Clarkia pulchella* (*pulchella* is Latin for beautiful). (The name *Lewisia* he saved for an equally interesting plant, bitterroot.) Over time it became apparent that there were many other, related species in the Far West, and nowadays over thirty species of *Clarkia* are recognized. They

Clarkia

show a remarkable range of flower form, some having tripartite petals, like ragged robin, others bipartite petals (that is, with two lobes instead of three), others paddle-shaped petals, still others simple petals rather like primroses.

A century and a half after its discovery by a tattered band of explorers, *Clarkia* has proved a challenge to botanists. Some species occur in broad ranges in the Pacific states, others in small pockets adjacent to the ranges of more widespread species. All can be grown under cultivation, a fact exploited by Harlan Lewis of the University of California in Los Angeles and others to determine the possible patterns of evolution within this curious group of plants. Major evidence was supplied by study of the chromosomes.

According to Lewis, the basic chromosome number in the genus is 7, but some species have 14, others 8, 9, 12, 17, or 26. This suggests that some may have arisen by a doubling of the chromosomes or by some other modification or multiplication. The more derived species, that is, those with the more unusual chromosome configurations, are generally those with more restricted ranges and unusual flowers; generally these occur in drier situations. There is a real possibility that each of these species was derived directly from one of the more widespread species by a sudden chromosomal change that better adapted it to a more rigorous environment. Because of the chromosomal differences, the species are not able to hybridize or at least to produce fertile offspring. Biologists refer to this as the "cataclysmic" origin of new species. That is, a species is formed abruptly through a chance modification of the chromosomes rather than very gradually through natural selection working on the slow accumulation of mutations in the genes that comprise the chromosomes. Such events may occur (primarily in plants rather than animals) when populations are subjected to severe drought in which most individuals are eliminated, but a few "aberrant" ones survive. Natural selection is still operating, but it is abrupt rather than slow.

William Clark (1770–1838), a man with little formal education and with a military background, would surely have been bewildered by this sophisticated research. His older brother, George Rogers Clark, was a general noted for his victories in the Indian wars following the Revolution, and William came to serve as a lieutenant in similar engagements under General "Mad Anthony" Wayne. For a time Meriwether Lewis (1774–1809) served in the same division, and the two became friends. Clark tired of the army (and of General Wayne) and in 1796 resigned and returned to his Kentucky home. In 1803 he received a letter from Lewis, who proposed that Clark accompany him on an expedition being sent by President Jefferson to explore the newly acquired Louisiana Territory all the way to the Pacific. Clark was enthusiastic and immediately began to make plans for the trip, even though he recognized it as "an immense undertaking fraited with numerous difficulties." He rounded up several of Kentucky's best hunters and woodsmen and joined Lewis near St. Louis to begin the organization of the expedition during the winter of 1803–1804.

Lewis was a Virginian who had (in the words of a fellow Virginian, Thomas Jefferson) "a talent for observation which had led him to an accurate knowledge of the plants and animals of his own country." Within a week after his election to the presidency, Jefferson had offered Lewis the position of his personal secretary, and over the next two years Lewis was carefully groomed for a project already in the back of Jefferson's mind: an exploration deep into western lands not formally acquired from France until 1803. Lewis was sent to Philadelphia for a brief course in medicine from Dr. Benjamin Rush, the most noted physician of the day, and for botanical training under Benjamin Smith Barton, whom we met briefly in the preceding chapter.

With advice from many others, Lewis amassed over a ton of supplies, which were taken by wagon to Pittsburgh and shipped down the Ohio by boat. The story of the Lewis and Clark Expedition has been told many times and needs no retelling here. Although less well educated than Lewis, Clark had more frontier experience, and he more than once saved the expedition from near disaster. He became an excellent cartographer as well as a naturalist of no mean ability. He was diligent in collecting specimens and drew careful sketches of birds, fish, and mammals. He kept a diary of each day's experiences, as requested by Jefferson (Lewis apparently failed to do so as faithfully). Clark's casual approach to spelling is sometimes amusing to us today; he wrote of musquetoes, whiper wills, and parrot queets. Yet there is no doubt that he responded as fully to details of his surroundings as did the better-educated Lewis.

The explorers discovered many colonies of prairie dogs as they crossed the plains. "The village [wrote Clark] . . . Contains great numbers of holes on top of which those little animals Set erect [and] make a Whisteling noise and whin allarmed Step into their hole." Clark called them ground rats, Lewis barking squirrels; it was another member of the expedition, John Ordway, who first began calling them prairie dogs (though of course they are not dogs, or rats, or squirrels).

Among the specimens shipped to Jefferson on the outgoing trip were six live animals: a prairie dog, a sharp-tailed grouse, and four magpies. The shipment left Fort Mandan (in present North Dakota) in a keel boat, then after several weeks in St. Louis was sent down the Mississippi to New Orleans, where all arrived alive. At New Orleans they were placed aboard the *Comet*, headed for Baltimore. Only the prairie dog and one magpie survived the long trip by sea, but they finally reached Washington after a trip of more than four thousand miles covering a period of nearly six months, with variable care en route. Jefferson was at Monticello, and it was another six weeks before he examined the animals. He then forwarded them to Peale's Museum in Philadelphia, at that time located in Independence Hall. Both survived for several more months, and when they eventually died they were stuffed and mounted. The magpie served as a model for Alexander Wilson's drawing for his *American Ornithology*, and the prairie dog for George Ord's description of the animal as a newly discovered species. Doubtless it is the only time that two living, wild creatures have ever inhabited both the Executive Mansion and Independence Hall.

Fortunately the natural history of the Lewis and Clark Expedition has been compiled in a detailed but entertaining way by Paul Russell Cutright in his book *Lewis and Clark: Pioneering Naturalists* (the source of the above anecdote). The expedition returned in September 1806, and the explorers made plans to publish their journals. But Lewis was shortly appointed governor of Louisiana Territory and Clark commander of the militia and superintendent of Indian affairs in St. Louis. With Lewis's untimely death in 1809, Clark and Jefferson persuaded Nicholas Biddle, a Philadelphia lawyer, to prepare the journals for the press. They were published in 1814 but with most of the observations on natural history omitted. It was not until 1893 that Elliott Coues published a more complete version of the journals, and not until 1904 that R. G. Thwaites published them in full, in several volumes.

Meriwether Lewis's death remains a mystery to this day. As governor of Louisiana, he found himself unable to cope with the necessary politics and red tape, and when his financial records were questioned he headed for Washington to defend himself. Traveling through Tennessee with two servants, he stopped at an inn, where he died in the night of bullet wounds. The official conclusion was that it was a suicide, but there was no money on his body and his watch was later found in New Orleans. So murder seemed more likely to Lewis's friends. Ornithologist Alexander Wilson had known Lewis in Philadelphia and went out of his way to visit the scene of Lewis's death and talk to those who were present at the time. His vivid account suggests strongly that Lewis was murdered, presumably for his watch and money.

William Clark spent most of the remainder of his life in St. Louis. In 1808 he married his childhood sweetheart, Judy Hancock, for whom he had named the Judith River in Montana. During the War of 1812 he was involved in several engagements with the English and with the Native Americans. He held his position as superintendent of Indian affairs for many years, and often escorted Native American chiefs to Washington and pleaded their causes. His home became a center of hospitality for both Native Americans and Whites.

Although the Lewis and Clark journals were not published in full until nearly a century after the expedition's return, many of the specimens they collected found their way to specialists fairly promptly. Alexander Wilson described some of the birds, including Clark's nutcracker and Lewis's woodpecker. George Ord described several of the mammals, including such notable western animals as the grizzly bear and the pronghorn antelope. Most of the plants found their way to Frederick Pursh (1774–1820). Pursh has sometimes been maligned, especially by Thomas Nuttall, who accused him of taking some of the plants Nuttall himself had collected and hoped to describe. John Bradbury, who traveled up the Missouri with a party of John Jacob Astor's American Fur Company, found on his return home that "this man [Pursh] has been suffered to examine the collection of specimens which I sent to Liverpool, and to describe almost the whole, thereby depriving me both of the credit and profit of what was justly due to me."

Pursh was born in Siberia and trained in Germany, but he moved to America in 1799, where he was employed at a botanical garden near Philadelphia and became acquainted with local naturalists William Bartram, Benjamin Smith Barton, and others. He took a number of collecting trips and soon began to assemble plant specimens from all over the continent. In 1811, for unexplained reasons, he suddenly left for England, taking with him all his collections, including those of Lewis and Clark, Nuttall, Bradbury, and others. In 1814 he published his *Flora Americae Septentrionalis*, the first *Flora* that attempted to cover all of America north of Mexico. In it he named many plants, including *Clarkia* and *Lewisia*, the latter often called bitterroot and the origin of the name for the Bitterroot Mountains and the Bitterroot River. The Native Americans dug out the roots, peeled them, and boiled them to remove the bitter flavor. Mountain men also enjoyed the roots as a supplement to a diet of flesh. Bitterroot is a short-stemmed plant with handsome pink blossoms, now the state flower of Montana.

In England, Pursh scandalized his colleagues by marrying a barmaid. Then he abruptly moved to Canada, where he died at the age of forty-five after a fire had destroyed most of his possessions. *Purshia* (bitterbrush) was named for him by Swiss botanist Augustin de Candolle in recognition of his contributions to botany. Bitterbrush is an abundant shrub on dry hillsides in parts of the West; it is a favorite browse of deer and elk and in the spring produces masses of pale yellow blossoms. It will do as a reminder of a dedicated if somewhat enigmatic botanist.

The plant specimens from the Lewis and Clark Expedition have had a curious history. When Pursh finished with them, he left them with a London patron, amateur botanist Aylmer Bourke Lambert (1761–1842). In his *Flora*, Pursh had named *Oxytropis lambertii* for him. Now called Lambert's loco weed, it is despised by ranchers because of its toxic properties and is a dubious tribute to a man who had helped him in many ways. In 1842 Lambert's herbarium was sold at auction, and an anonymous American purchased the specimens and returned them to Philadelphia. Half a century later, construction workers in the building of the American Philosophical Society discovered a forgotten bundle of plants in the basement. They proved to be the long-lost plants collected by Lewis and Clark and described by Pursh. They were returned to the Academy of Natural Sciences, where the now much-faded specimens can still be seen by interested persons.

Claytonia

There is perhaps no more fitting memorial to any naturalist than to have a genus of plants named for him or her, especially if the plants are especially treasurable, as are the species of *Clarkia*. Linnaeus named one of the most treasured of all genera, spring beauty, the first flower of spring, for John Clayton (1694–1773), calling it *Claytonia*. Clayton came to Virginia from England in 1770 and began to collect plants and seeds and send them to his friend, London merchant Peter Collinson. Many found their way to Linnaeus and to Dutch botanist Johann Gronovius. Clayton prepared a catalog of Virginia plants, which he sent to Gronovius, who published the *Flora Virginica* under his own name after revising it according to the Linnaean system. Spring beauties of several species occur all across the continent, in places blooming as early as February. The small tubers are a favorite food of bears and were consumed by Native Americans either raw or cooked.

Clintonia

Clintonia, a fragile member of the lily family, was named for New York governor DeWitt Clinton (1769–1828) by Constan-

tine Rafinesque. One of the most brilliant and versatile of frontier naturalists and also one of the most eccentric, Rafinesque
will have a chapter to himself. Clinton was much more than a
politician. Through his newspaper column, *Letters on Natural
History and Internal Resources of New York*, he became one of
the most widely read naturalists of his time. From his position
of power, he was able to be of much help to Rafinesque,
Audubon, David Douglas, and other naturalists. Species of
Clintonia occur from coast to coast; sometimes they are called
queencup or beadlily, but more often simply *Clintonia*.

Cassin's Finch

John Cassin (1813–1869) was never honored by a generic name,
but he did have a journal named for him—*Cassinia*, the publication of the Delaware Valley Ornithological Club. Cassin was
brought up in Delaware County, Pennsylvania, and educated in
local Quaker schools. He became a successful businessman and
was soon attracted to Philadelphia's Academy of Natural Sciences. In his spare time, Cassin began to organize the
Academy's large collection of bird specimens. He received birds
collected on various expeditions then exploring many parts of
the world, and he soon came to know the world bird fauna as no
one else. He never held a salaried position at the Academy, but
it did publish many of his papers. Several young naturalists
owed their start to him, including Graceanna Lewis, one of the
few woman naturalists of the nineteenth century.

Species named for John Cassin include Cassin's auklet, a
bird that breeds on isolated cliffs along the coast of Alaska;
Cassin's kingbird, a handsome flycatcher of the Southwest;
Cassin's sparrow, another species of the semiarid Southwest,
where it lives among cacti and mesquite; and Cassin's finch,

widely distributed in the West. In the Rockies, Cassin's finches (*Carpodacus cassinii*) are common at winter feeders, and it is a special pleasure to watch the red cap and throat of the males intensify as spring approaches. In summer they move higher in the mountains and build their nests in Douglas-firs and other conifers. Their song is among the most melodious to be heard in alpine forests.

Comstock's Mealybug

Only an entomologist would appreciate having a mealybug named for him. Surely John Henry Comstock (1849–1931) would not have been in the least distressed, since he had devoted some of the best years of his life to the classification of mealybugs and their relatives, the scale insects. These are minute creatures, covered with waxy or fluffy excretions that fasten themselves tightly to plants and suck their juices. They have little obvious body form, but Comstock discovered that they could be classified by their pygidia (rear ends). His wife Anna set to work to draw the tail ends of scale insects and mealybugs, surely the ultimate in wifely devotion. Comstock's report was published by the U.S. Department of Agriculture in 1881.

Comstock was, at the time, on leave from Cornell University, where he had begun teaching five years earlier. Cornell's Department of Entomology was the first in the country, and it attracted many students who went on to distinguished careers. Comstock's book *An Introduction to Entomology*, which went through several editions, attracted many to this field (including this writer, though some years after Comstock's death). Comstock had more than a dozen insects named for him, but he seems fated to be best remembered for Comstock's mealybug

(*Pseudococcus comstocki*), since it is a notorious pest. Although fewer than 2 percent of insect species can be classed as serious pests, it is the pest species that have been given common names, since they must be dealt with by society. To invent common names for the other 98 percent of the million or so species of insects would be a hopeless task. However, that most are known only by their scientific names is a handicap to entomologists when they try to communicate with the public.

Dall's Porpoise

*D*olphins and porpoises are much in the news, since their numbers are declining as the result of their drowning in nets placed to catch tuna. The Japanese harvest them as food now that whaling is subject to international regulation. Some populations are also being decimated by mysterious diseases, perhaps induced by PCBs and other toxins we are dumping into the ocean. The seas would be dreary indeed without these sleek and exuberant animals leaping in schools behind ships. In captivity they can be taught to respond to a number of human signals, and nowadays no Seaworld or Marineland would be complete without a few performing dolphins.

Dall's porpoise (*Phocoenoides dalli*) is not one of the better known of its clan, since it doesn't do well in captivity. But it is the most strikingly colored of all porpoises, jet black with a large white patch on its belly and lower sides, usually also a bit of white on its dorsal fin and on its small tail flukes. These are denizens of the North Pacific, ranging all the way from the Bering Sea to the coast of California. Although they do ride the bow or stern of ships, they do not usually leap from the sea in the manner of other porpoises, but rush to the surface sending a "rooster tail" of spray. They have a robust body, at their maximum size about seven feet long and weighing about five hundred pounds. Like most porpoises they have a blunt head, without the "bottle nose" of the more familiar dolphins.

Much has been written about the supposed intelligence of dolphins and porpoises, particularly following John C. Lilly's 1967 book *The Mind of the Dolphin: A Nonhuman Intelligence*. It is true that the brain of an average-sized dolphin is comparable in size and convolutions to that of the human brain, and there may be more going on in that brain than we realize. That dolphins cooperate to assist injured individuals is well known, and females sometimes help newborn calves that are not their own. Dolphins also readily imitate sounds and actions and of course can be taught a variety of tricks. But if these oceanic mammals have an "alien intelligence," as is sometimes claimed, it has yet to be established.

Dolphins, porpoises, and whales do communicate, using a variety of whistles and sometimes eerie sounds that carry great distances through the water. They also use ultrasonic clicks, a form of sonar, to locate prey. The whistles are produced by the larynx, like most mammal sounds. The clicks, however, are pro-

Dall's Porpoise

duced by powerful muscles associated with the nasal passages and are not passed out through the blow-hole but are discharged directly through the enlarged forehead, after passing through a mass of fatty tissue. These are intense, patterned, ultrasonic pulses that are reflected back from their prey. But since the prey fishes can sometimes swim faster than the predator, it has often been asked whether the ultrasound may not stun the prey. There is no solid evidence that stunning occurs, but at least the prey are sometimes disoriented, and that may be enough to permit their capture. Research on this subject will go on for many years. These remarkable animals deserve our fullest respect and should never be killed recklessly in fish harvests or as food when meat from domestic animals is available.

Dall's porpoise is a relatively silent species, apparently lacking the communicative whistles of many other species, perhaps because it travels in groups of ten to twenty rather than in larger groups. Calves are born in the summer after a gestation period of nearly a year; they are nursed by their mothers for about two years. They eventually develop a set of sharp teeth and join their elders in the hunt for squid, fish, and mollusks, often diving deeply to find them. Although Dall's porpoises have been considered to be fairly common in the North Pacific, no one knows how many there really are.

William H. Dall (1845–1927) discovered the porpoises off the coast of Alaska in 1873. Through the influence of Spencer Baird, Dall had been appointed to the U.S. Coastal Survey, and he was working the Alaska coast and the Aleutians. He was already a veteran of the Far North, having been a member of an expedition launched by the Western Union Company in 1865 to determine the feasibility of establishing a telegraph line to Europe through Alaska and Siberia (the first Atlantic cable had recently failed). When the leader of the expedition, Robert Kennicott, died suddenly, Dall took command of the expedition, although he was only twenty-one. Following the 1865 expedition, he published *Alaska and Its Resources*, for many years a major source book on that then little-known part of North America.

Dall was a Bostonian, son of a Unitarian minister of some renown. He developed an interest in natural history when he was very young and began a collection of New England shells. Working for a time as a clerk on India wharf in Boston, he used every idle moment to copy scientific books he thought he would never be able to buy. Later he came under the influence of Louis Agassiz at Harvard and of Robert Kennicott at the Chicago Academy of Sciences. In 1880 he was appointed honorary curator at the U.S. National Museum, where he specialized in mollusks. In a room in a tower of the Smithsonian, he wrote hundreds of scientific papers, many on mollusks of the West Coast, including some of those collected by Edward Palmer.

Dall was a pioneer in the study of ecophenotypes, that is, the different forms that a species may assume in different environments. His study of the different shapes of oyster shells, depending upon substrate and population density, has been cited many times and served as an antidote to the tendency of C. Hart Merriam and others who frequently described slight variants as new species. (Merriam, Palmer, and Kennicott will each make an appearance in later chapters.)

Dall also made substantial contributions toward the standardization of the rules for the naming of animals. Other publications of his dealt with topics in anthropology, paleontology, tidal currents, and meteorology. Following Baird's death in 1887, he wrote a biography of his mentor. He was indeed a man of many talents. Besides the porpoise, Dall's sheep, Dall's limpet, and several other animals are named for him.

Dall sent his notes and specimens to the Smithsonian Institution, where most of the mammals came to the attention of the curator of mammals at the National Museum, Frederick True (1858–1914). True was a native of Connecticut who was

trained at City University of New York and entered government
service as a clerk. He was promoted to librarian of the U.S.
National Museum and in 1883 became curator of mammals;
later he assumed various administrative duties at the Smithsonian. True was a quiet and studious person, fond of music and
good literature but impatient with social activities that diverted
him from his research. Because of his poor eyesight he turned
his attention to the very largest of mammals, the whales, dolphins, and porpoises, and became an expert on both fossil and
living marine mammals. In the course of these studies he
named Dall's porpoise as well as several other oceanic mammals. He traveled frequently to other museums and to whaling
stations, but he was not really a field naturalist. The pinyon
mouse, *Peromyscus truei*, and the Sonoran squirrel, *Sciurus
truei*, are named for him. It is ironic that two small land mammals are named for a person who preferred them wet and large.

Douglas-fir

In a book of eponyms, it would not do to pass up David Douglas (1798–1834), who is said to have had more plants named
for him than anyone else in the history of botany. One thinks
immediately of Douglas-firs, which surpass all other North
American tree species in the value of the lumber produced.
Exploring Oregon Territory in 1825, Douglas was awed by
these great trees; one that he measured was 227 feet tall and 48
feet in circumference. Douglas-firs are not true firs, nor are
they spruces, though they look a bit like both. The cones are
unique: they are pendant, with bracts extending between the
scales; each bract has a double end with a slender filament
between, rather like the tail end of a tiny animal plunging into
the cone.

Douglas was not the first to discover these trees. That distinction belongs to Archibald Menzies, for whom the trees are named scientifically (*Pseudotsuga menziesii*) (more about Menzies in a later chapter). But Douglas was the first to ship seeds back to England. He was employed by the Royal Horticultural Society of London to find plants in the Pacific Northwest that would grow well in English parks and gardens. A great many were dispatched to England, not only trees but flowering plants such as *Clarkia*, California poppy, and mountain pink (*Douglasia nivalis*). He also discovered many plants for the first time, including some of the Far West's most distinctive trees: sugar pine, western white pine, silver fir, Oregon white oak, and others.

Identification of conifers depends in considerable part on the cones, and Douglas was often frustrated by the size of these trees in the virgin forests, where the cones were far out of reach. He told in his diary how he had a narrow escape collecting cones of sugar pine. "Being unable to climb or hew down any [trees], I took my gun and was busy clipping [cones] from the branches with ball when eight Indians came at the report of my gun. They were all painted with red earth, armed with bows, arrows, spears of bone, and flint knives, and seemed to me anything but friendly."

Douglas offered them tobacco if they would collect cones for him, and as soon as they left to look for cones, he slipped away. Douglas had learned the Chinook language and generally got on well with the Native Americans, who thought he was crazy. He was often alone as he roamed the wilds, with little protection from the elements or from insects; often he lived for weeks on berries or whatever game he could shoot.

At Fort Assiniboine, in western Canada, he met John Franklin, who had just returned from the Arctic with other members of his expedition, including John Richardson and Thomas Drummond, the latter also employed by the Royal Horticultural Society. In 1827 Douglas crossed the Rockies to Hudson's Bay, then returned to England. But he seemed ill at ease away from the wilderness and was soon sent out again, this time to California. Here he visited groves of coastal redwoods and collected, among other things, several new species of mariposa lilies. From California he went to Hawaii briefly, then back to the Columbia River and on into British Columbia. Here he lost most of his possessions, and nearly his life, when his canoe plunged over a cataract. He wrote to a friend: "When I tell you that I was an hour and 40 minutes in the water in the rapids myself and after escaping had 300 miles over a barbarous country without food or shelter you will form an idea of my condition. . . ."

In 1833 Douglas resigned from the Horticultural Society in a fit of pique and once again sailed for Hawaii. Here he climbed all three mountains on the island of Hawaii—Mauna Loa, Mauna Kea, and Kilauea. Then one day he fell into a trap set to capture wild cattle and was trampled to death by a bull. It was rumored that the man who had built the trap had pushed him in because Douglas was having an affair with his wife. The possibility of murder was investigated but rejected for lack of evidence. Douglas had become very nearsighted, and it is much more probable that he simply did not see the trap. He was only thirty-four when he died.

Besides Douglas-fir and *Douglasia*, many species have been named for him, including Douglas's hackberry, *Celtis douglasii*; water-hemlock, *Cicuta douglasii*; and Douglas's knotweed, *Polygonum douglasii*.

Drummond's Mosses

Douglas apparently got on well with Thomas Drummond (?1790–1835) during their brief time together. Drummond had been the botanist on Franklin's second expedition to the Arctic coast. He had not gone all the way to the coast but had spent several months, traveling alone, in central Canada. At times he suffered from bitter cold and near-starvation. Nevertheless he collected some 1,500 plant specimens, 150 birds, and about 50 mammals. At York Factory, on Hudson's Bay, Douglas and Drummond, with three others, set out in a small boat to visit the ship *Prince of Wales*, anchored offshore. A fierce wind came up and blew them sixty to seventy miles out in the bay, and it was not until the next day that they were able to row themselves, with a favorable tail wind, back to shore. Both men seemed fated to live lives of desperate adventure that ended tragically, but not at that moment.

Like Douglas, Drummond was a Scot, and like him he had a reckless drive to collect everything in sight for his British patrons. Drummond returned to Scotland from Canada in 1827 and spent two years as curator of the Botanical Gardens in Belfast. Here he published *Musci Americani*, a treatment of over two hundred kinds of mosses he had collected on his trip with Franklin. "The whole continent of North America [wrote Glasgow botanist William Jackson Hooker] has not been known to possess so many mosses as Mr. Drummond has detected in this single journey."

In 1830 Drummond was off to America again as an emissary of Hooker. After visiting John Torrey in New York, he went to New Orleans and then to Texas, which had been sparsely settled by Americans but was still legally part of Mexico. This was virgin country for a naturalist, and Drummond shipped back many new plants as well as several birds and snakes. Hooker wanted him to go as far as Santa Fe, but Drummond pointed out the distances involved and the threat of Indian attack. Cholera was rampant in Texas, and Drummond spent several weeks too ill and weak to collect specimens. He had scarcely recovered from cholera when he wrote to Hooker in his last letter from Texas that he had "such a breaking out of ulcers, that I am almost like Job, smitten with boils from head to foot."

Drummond decided to return home, then bring his family to Texas (oddly, considering his sufferings!). Here he could buy "a league of land for one hundred and fifty dollars" and roam widely, perhaps even into Mexico. He went first to Florida, then on to Havana, where he died a mysterious death, probably of disease. He was about forty-five years of age and died two years after Douglas. Hooker wrote: "Thus have perished, while engaged in the cause of science with a zeal of which history presents few examples and nearly at the same time, two men in the prime of life. . . ."

Of the flowering plants named for Drummond may be mentioned Drummond's rock cress, *Arabis drummondii*; Drummond's milk vetch, *Astragalus drummondii*; and *Phlox drummondii*. The last proved to do well in cultivation and is now widely grown in gardens. Drummond collected many mosses in Texas and several of these also bear his name. Two subspecies of mammals were also named for him on the basis of specimens he took in Alberta.

Engelmann Spruce

*S*pruces are the most abundant of all North American trees, sweeping in dense stands all across Canada and the northern tier of states and well down the spine of the Rockies, flourishing in places too cold and damp for much human settlement. To most of us, a spruce is a spruce, an evergreen with short, prickly needles and smallish, pendant cones. Rich green and fragrant, these are fine ornamentals and ideal Christmas trees. In December 1990 the small city of Walden, Colorado, shipped a seventy-five-foot Engelmann spruce to the White House to serve as the national Christmas tree. It left Walden to a parade and was accompanied all the way to Washington by a contingent of Waldenites.

Distinguishing between the half dozen native kinds of spruces and several imports is not an easy matter. George Engelmann, a St. Louis physician, was fascinated by conifers and spent several summers in the field studying them and delineating the various species. It is fitting that one of the handsomest and most notable trees of high altitudes in the Rockies be named for him: *Picea engelmannii*.

Travelers above nine thousand feet altitude in the southern Rockies, or considerably lower in the northern Rockies, are invariably impressed by the towering symmetry of these trees and the density of their stands. Often the trees attain over one hundred feet in height, with a trunk diameter of three feet or more. In spite of the trees' size, the cones are only two or three inches long, clustered near the top of the tree, perhaps an adaptation serving to reduce self-pollination. Trees may live for several hundred years unless they are consumed by fire or fall to the axe. Old trees fall to form moss-covered stumps and

Engelmann Spruce

logs that decay over many years, providing habitat for beetles and rodents before becoming part of the forest soil. To enter a forest of Engelmann spruce is to enter a primeval world, dark and damp, with few wildflowers and hardly a bird song to be heard.

These forests are frost-free for no more than two or three months of the year. The soil soaks up and holds moisture like a sponge, not only moisture that falls on the forest itself but much of that draining the tundra above. Spruce forests may go on for many miles within their altitudinal range, pure stands except for an intermingling of subalpine fir. Young trees are able to take root in the shade of the forest, ready to grow rapidly and take the place of fallen giants. At their lower limits, the spruce may be mixed with aspens and lodgepole pines, while at their upper limits the trees are small and distorted by the wind and by avalanches, the twisted trees of timberline.

George Engelmann (1809–1884) was born in Germany and trained there in medicine, though from youth he was absorbed in the world of plants. For a short time he studied with the young Louis Agassiz; in 1832 he sailed for America, more than a decade before Agassiz made a similar move. Engelmann settled in St. Louis at a time when that city was growing rapidly as a gateway to the West. His practice was soon flourishing, but he found time, between patients, to slip into a back room where he kept his library and herbarium. In 1840 he visited Asa Gray at Harvard and agreed to find collectors in the West who would forward plants to Gray, for a price. Over the next few years nearly every naturalist traveling in the West visited Engelmann, who advised them on their itinerary and sometimes advanced them money, to be repaid when they were reimbursed for their specimens. By 1869 Engelmann was seeing fewer patients and devoting most of his time to botany. He made several trips through the West to study conifers. On one trip, he stopped at the Great Salt Lake to take a swim, "floating like a buoy and providing some amusement for others due to his portly dimensions," according to Joseph Ewan in his book *Rocky Mountain Naturalists*.

Engelmann published several monographs and is also credited with discovering the role of the yucca moth in the pollination of yuccas. He also discovered that American grapes were resistant to the attacks of an aphid called the grape phylloxera. As a result of this discovery, European wine producers were able to graft their vines onto American rootstocks and so avoid devastation of their vineyards by this pest. Engelmann's herbarium eventually became the nucleus of the Missouri Botanical Gardens.

Among the trees Engelmann described were bristlecone pine, whitebark pine, pinyon pine, lodgepole pine, and Colorado blue spruce. He also described two of the Southwest's most remarkable plants, the saguaro cactus and the Joshua tree. Besides the spruce, Apache pine (*Pinus engelmannii*), Engelmann oak (*Quercus engelmannii*), a daisy (*Engelmannia pinnatifida*), and several other plants were named for him. Engelmann Peak, in the Colorado Rockies, stands not far from peaks named for Asa Gray, John Torrey, and Edwin James, all notable botanists.

It was Englemann's good friend and associate Charles Parry (1823–1890) who named both Engelmann Peak and Engelmann spruce. Like Engelmann, Parry was trained in medicine but attracted to plants from his youth. At Columbia, where he took his M.D. degree, he became acquainted with John Torrey. For a few years, Parry practiced medicine in Iowa, but the lure of botany was great, and in 1849 he was appointed botanist of the United States and Mexican Boundary Survey. His report, written jointly with Torrey, was one of the first accounts of the unusual flora of the deep Southwest. For the next forty years, Parry continued to live in Iowa but spent his summers exploring the West for plants, either on his own or as botanist to some exploring party. He built a cabin on Grizzly Gulch, west of Denver, where Asa Gray, John Torrey, and other naturalists visited him from time to time. On a long trip to the forests of the Pacific Slope, his companion was George Engelmann. Late in his life, Parry spent much time in California studying the flora of the chaparral zone.

The many newspaper articles Parry wrote described some of his travels vividly. Unfortunately he had little patience with paperwork and did not keep careful records on the specimens he collected. His three years at the Smithsonian Institution (1869–1871) ended when he found himself unable to cope with governmental red tape. Above all he was an explorer for plants, a gentle, friendly man who roamed in places that were occupied (if at all) by rough miners and woodsmen. The number of plants named for him is legion: Parry's primrose, Parry's clover, Parry's gentian, Parry's harebell, and many others. Parry, too, had a peak named for him, by Surveyor General F. M. Case. In the words of a biographer, "his happy personality is . . . associated with the most romantic and fruitful period of botanical exploration in the Far West."

Eaton's Grasses

The botanical circle that included Engelmann, Torrey, Gray, and Parry owed much to a naturalist of the previous generation, Amos Eaton (1776–1842), whose *Manual of Botany for the Northern States* went through eight editions between 1817 and 1840. Eaton was trained in law, but early in his career he found himself imprisoned for life on a trumped-up charge of forgery. While in prison, he planned a classification of plants based on Linnaeus's system of counting stamens and other flower parts. He began to teach botany to the prison inspector's young son,

John Torrey. After four years, Eaton was released, but he was exiled from his home state, New York. Soon after, however, a new governor, DeWitt Clinton, not only pardoned him but invited him to give a series of lectures in Albany.

Eaton went on to study at Yale and to teach at his alma mater, Williams College. He was a big man, with a high forehead, and evidently a brilliant speaker. At Williams, his courses were so popular that (in his words) "an uncontrollable enthusiasm for Natural History took possession of every mind; and other departments of learning were, for a time, crowded out of College." One of his students was the poet William Cullen Bryant, an ardent amateur botanist. "I affected to be your superior [Eaton wrote to him some time later] because I knew the names of more weeds than you."

Eaton's lectures were couched in simple, everyday words, and he became the leading popularizer of natural history of his day. At Williams and later at the Rensselaer Institute, he often took his students on field trips. Once, on a trip down the Erie Canal, he was joined by a "curious Frenchman," Constantine Rafinesque, who later spoke of the trip as "one of the most agreeable journeys I ever performed." One of Eaton's students was Almira Hart Lincoln, an unusual woman whose *Familiar Lectures in Botany* became a standard text for female seminaries. Mrs. Lincoln, like Eaton, spoke of the parts of flowers while carefully avoiding mentioning that they were sexual organs analogous to those of animals.

Young Asa Gray carried Eaton's *Manual of Botany* with him as he roamed the fields of upstate New York in the search for plants. But Torrey and Gray went well beyond Eaton, who stuck to Linnaeus's "sexual system" through all the editions of his book, even though this often resulted in the grouping of very dissimilar plants simply because they had the same number of stamens. Torrey was the first to use more broad-based criteria for grouping plants, a "natural system," leading Eaton to say of one of Torrey's publications, "No book has probably excited such consternation and dismay."

Eaton's success as a lecturer drew the attention of Stephen van Rensselaer, who engaged him to help survey the country through which the Erie Canal passed. As a by-product of the survey, Eaton produced a pioneering study of the geology of the northeastern states. He was involved in the founding of the American Geological Society in 1819. When Rensselaer Polytechnic Institute was founded in 1824, Eaton became one of its major professors. He was primarily a teacher and popularizer and made few contributions to botanical research. Perhaps for that reason, few plants have been named for him. Rafinesque did remember him, however, naming a genus of grasses *Eatonia*. The name has fallen into disuse, but these grasses are still called by the common name Eaton's grasses.

Eaton's Ferns

Amos Eaton's grandson, Daniel Cady Eaton (1834–1895), earned a very substantial reputation as a student of ferns. After working for a time as a student of Asa Gray at Harvard and after a stint in the Civil War, he joined Clarence King's geological survey of California, permitting him to make extensive collections in the West. On his return he was elected to a newly established professorship of botany at Yale. He remained there for many years, producing monographs on ferns based on his own collecting and on specimens collected by various surveys of the West. His two-volume book *The Ferns of North America* was well illustrated and covered all the ferns then known from north of Mexico. Eaton's lipfern, *Cheilanthes eatoni*, bears his name.

Eschscholzia *(California Poppy)*

In contrast to the Eatons, Johann Friedrich Eschscholtz (1793–1831), though he spent only a few weeks in North America, left his name on quite a number of plants and animals, including one of the most admired of plants, the California poppy. His story begins far from the Eatons', in Russia. In 1814 the Grand Chancellor of Russia, Count Romanoff, planned an expedition to explore the West Coast of North America, parts of which were claimed by Russia. The ship was the *Rurik*, commanded by Otto von Kotzebue and carrying two naturalists, Eschscholtz and Adelbert von Chamisso. Leaving St. Petersburg in May 1815, the ship rounded Cape Horn and arrived off the coast of California (by way of Kamchatka) in October 1816. At San Francisco, they were welcomed cautiously by the Spanish, and the naturalists were permitted several days in the environs. "The Flora of this country is poor," wrote Chamisso. Nevertheless he collected and later described one of the most brilliantly colored of all flowers and named it for his colleague, calling it *Eschscholzia californica*. One might have hoped for a more euphonious name for the California poppy, but at least it is a way of remembering an otherwise mostly forgotten naturalist. It may seem odd that the genus named for Eschscholtz omits the *t* in his name. In fact, when he moved from his native Germany to Russia he dropped the *t*, and Chamisso accepted this spelling.

Eschscholtz accompanied Kotzebue on a second trip to California, in 1823–1826, and this time he was able to stay several weeks and to range a bit farther from San Francisco. Animals collected during these weeks and later described by Eschscholtz include the Pacific giant salamander, the California slender salamander, the Pacific sand dollar, and several beetles. Plants he described include coffeeberry, blue bush, yellow sand verbena, and Chamisso's lupine. On a brief stop in the Hawaiian

Islands, on his first trip, Eschscholtz discovered one of only two butterflies native to the islands, which he named *Vanessa tameamea*, the Kamehameha butterfly. He also made significant collections of plants and animals in Alaska, the Aleutian Islands, the Philippines, and Chile. After his return, he was made a professor at the University of Dorpat, Estonia, and director of the university's museum. He held these positions for only a few years before he died at the early age of thirty-eight. In California, especially, his name is still very much alive.

Forster's Tern

Terns are contradictory creatures, at once the most graceful and ethereal of birds and at the same time harsh-voiced and tireless predators. To see a tern hovering head-down over the water, then plunging head first and emerging a moment later with a struggling fish in its stiletto-like beak, is to witness the actions of a superb hunter. The slender, pointed wings and deeply forked tails enable terns to twist and turn quickly as they scan the water for fish. But why several species are similarly colored—gray and white plumage, black caps, and red or orange bills and feet—is harder to explain unless one assumes all evolved from a common ancestor that somehow settled on this garb as most fitting for this mode of life. Whatever the reason, distinguishing the several species poses problems for bird-watchers. Forster's is the tern most often seen on inland waters of western North America, and in winter it is the one most often seen on our southern coasts. To most of us it is separable from the common tern and several other species only if we have a close view and quick access to a bird guide.

Forster's terns (*Sterna forsteri*) breed in marshes in the Great Basin and in western Canada. Unlike other terns, which usually

Forster's Tern

nest in close-packed colonies, often noisy and tumultuous, these terns nest in scattered pairs across marshes. Courting pairs fly up to a considerable height, with long, sweeping wing strokes, then glide down side by side on stiff wings. On the ground they weave back and forth, facing one another, with tails depressed and wings arched. The nests are fragile platforms built on hummocks or muskrat houses. Adults share in the incubation of the three grayish-brown eggs. At one time these beautiful birds were harvested as ornaments for ladies' hats. Nowadays their numbers are threatened as wetlands are taken away from them, one by one.

Johann Reinhold Forster (1729–1798) seems an odd person to have given his name to a North American bird, since he never visited the continent. He was born and trained in Germany, although his parents were of Scottish descent. He spent his early life in Danzig (now Gdansk, Poland). In 1765 he was commissioned by the Russian government to study the plight of the Germans who had settled on the Volga River. His report was less glowing than the authorities expected, and in response to criticism he decided to move to England. Here he began to correspond with Linnaeus and other naturalists. In Danzig he had prepared a German translation of Linnaeus's *Systema Naturae*, and in England he edited an English translation of Peter Kalm's *Travels in North America*.

After only three years in England, Forster was appointed naturalist on James Cook's second voyage to the Pacific. His oldest son, George, then eighteen, went along to collect and sketch specimens. The expedition went farther into the Antarctic than any ship had yet penetrated. Forster complained bitterly about conditions on board and about the lack of opportunity to make landings where he could study the fauna and flora. Nevertheless he made novel studies of albatrosses and petrels, and he discovered five new species of penguins. After the expedition's return, George Forster published an account of the voyage in which Cook was criticized to the point of libel. Johann Forster was a temperamental, irascible man, and some of his quarrelsome nature had evidently rubbed off on his son.

Cook's biographers have been most unkind to Johann Forster. Alan Villiers spoke of him as an "alleged scientist" and a "hack." "If Admiralty [Villiers wrote] had set out deliberately to inflict upon Cook and everybody else in the *Resolution* the most troublesome, useless, and energetically hostile a shipmate it was possible to find, they could not have made a better choice." J. C. Beaglehole, another biographer, called him a "problem from any angle," and "one of the Admiralty's vast mistakes."

Of course the British have never taken kindly to criticisms of one of their national heroes, James Cook. When Johann Forster returned to Germany, after the voyage, he became a professor at the University of Halle for eighteen years, where he published voluminously and created a very different impression. Young Alexander von Humboldt visited him before he took off on his own explorations, and spoke of how much he had gained from his publications. A biographer, J. G. Meusel, spoke of Forster's "vast learning" and said that if he ever met him in person he would "fall down" and "worship" him. Another called him "the Ulysses of these regions," meaning the South Seas. Perhaps Forster mellowed in his later years; or perhaps his German countrymen had learned to overlook his prickly disposition. George Forster, incidentally, became involved in the French Revolution and died in a Paris garret in 1794.

While in London in 1771 Johann Forster wrote an article about several birds sent to the Royal Society from Hudson's Bay. Among them were specimens he described as representing a variety of the common tern. Thomas Nuttall, when he prepared his *Manual of the Ornithology of the United States and Canada* (1832), recognized the variety as a distinct species, which he named *Sterna forsteri*, "from the eminent naturalist and voyager who first suggested these distinctions." In fact Lewis and Clark had discovered the tern in 1806 on their epic journey through the Northwest, but they did not give it a scientific name, and in any case their journals were not published in full until many years later.

When Thomas Nuttall (1786–1859) published his manual of ornithology he had already established himself as the leading botanist of his generation. His *Genera of North American Plants, and a Catalogue of the Species, to the Year 1817* had been published in 1818. Already he had traveled up the Missouri River with members of John Jacob Astor's American Fur Company and collected many new plants as well as insects for his friend Thomas Say. Washington Irving wrote of him (in *Astoria*): "Delighted with the treasures, he went groping and stumbling along among a wilderness of sweets, forgetful of everything but his immediate pursuit. The Canadian voyageurs used to make merry at his expense, regarding him as some whimsical kind of madman." On one occasion, the party's leaders checked all firearms in anticipation of an Indian raid, and found that Nuttall's gun was packed with dirt: he had been using it to dig plants! More than once he was hopelessly lost and had to be rescued. The Native Americans feared him as a powerful medicine man, but on one occasion rifled his possessions and drank most of the alcohol in which his specimens had been preserved.

Nuttall had been born in Yorkshire, England, and trained as a printer. Arriving in Philadelphia when he was twenty-two, he became acquainted with Benjamin Smith Barton, professor of botany at the University of Pennsylvania. Barton hired him to collect plants for him in the West—at eight dollars a month plus traveling expenses. In Michigan, Nuttall met up with the Astorians, who invited him to accompany them up the Missouri River. Here his companions, from time to time, were two other young men eager to explore this newly opened country: John Bradbury and Henry Brackenridge. Both published accounts of their travels.

In 1819 Nuttall went off on his own, exploring the lower Arkansas Valley, then virgin territory. This trip, too, was filled with mishaps and dangers. He became ill and for a time had to be helped on and off his horse by a trapper friend who accompanied him. At times there was little to eat but beaver tails and wild honey. There were botanical treasures, to be sure, but Nuttall "could not help indeed reflecting on the inhospitality of this pathless desert."

Nuttall wrote of his adventures in A *Journal of Travels into the Arkansa Territory, During the Year 1819.* Back in Philadel-

phia, he renewed his friendship with Thomas Say, who returned
in 1820 from his own trip down the Arkansas, as a member of
the Long Expedition. The two worked at describing their
discoveries and often gave lectures at the Academy of Natural
Sciences.

Then, in 1822, Nuttall received an unexpected opportunity:
he was offered a position as instructor of natural history and
curator of the botanical gardens at Harvard—in spite of the fact
that he had no academic background. He was paid five hundred
dollars a year plus the fees students paid to take his classes. He
was a popular figure and often took his students into the field.
Since there was no suitable textbook, he wrote his own, *Introduction to Systematic and Physiological Botany*. It was there
in Cambridge that he also wrote his *Manual of Ornithology*, a
field book and therefore more condensed than Wilson's or
Audubon's pioneering works. Audubon visited Nuttall, who
showed him where to find an olive-sided flycatcher, which
Audubon had never seen.

But Nuttall was restless at Harvard, and after a decade there
he left and joined the expedition of Nathaniel Wyeth to the
Pacific, where his companion was a young Philadelphia physician and naturalist, John Kirk Townsend. (Nuttall later named
a handsome genus of composites *Wyethia*.) Townsend's book,
Across the Rockies to the Columbia, is filled with stories of their
adventures as they traversed this pristine landscape. As they
approached the Pacific they were racked by violent storms, and
Nuttall's "large and beautiful collection of new and rare plants"
became thoroughly soaked, so that the specimens had to be
dried if they were to survive.

In this task [wrote Townsend] he exhibits a degree of
patience and perseverance which is truly astonishing; sitting
on the ground, and steaming over the enormous fire, for
hours together, drying the papers, and re-arranging the

whole collection, specimen by specimen, while the great drops of perspiration roll unheeded from his brow. Throughout the whole of our long journey, I have had constantly to admire the ardor and perfect indefatigability with which he has devoted himself to the grand object of his tour. No difficulty, no danger, no fatigue has ever daunted him, and he finds his rich reward in the addition of nearly *a thousand* new species of American plants, which he has been enabled to make to the already teeming flora of our vast continent.

Once on the coast, Nuttall traveled to Hawaii and then to California. There, on a beach, he was found by one of his former Harvard students, Richard Henry Dana, stuffing shells into his bulging pockets, an event recorded in *Two Years Before the Mast*. Nuttall returned east via Cape Horn. Back in Boston, he worked on his collections and gave several well-attended lectures. Then, at the age of fifty-six, he went back to England, where he had inherited an estate. He returned to Philadelphia briefly in 1847, but otherwise he remained at his home near Liverpool. It is said that he died soon after overexerting himself while opening a box of plants he had just received.

Nuttall's first biographer, Elias Durand, described him as "a remarkable-looking man: his head was very large, bald, his forehead expansive, features diminutive with a small nose, thin lips and round chin, height above middle, person stout with a slight stoop and his walk peculiar and mincing, resembling that of an Indian."

In a mere thirty-four years Nuttall had become the most well-traveled and knowledgeable of all American naturalists. Many plants and animals are named for him: among them are Nuttall's woodpecker, Nuttall's blister beetle, Nuttall's sunflower, and Nuttall's evening primrose. Audubon named the yellow-billed magpie *Pica nuttalli*, remarking that "I have conferred on this beautiful bird the name of a most zealous, learned and enterprising naturalist, my friend Thomas Nuttall, Esq., to whom the scientific world is deeply indebted for many additions to our zoological and botanical knowledge which have resulted from his labors." It is difficult to roam very far in the West without discovering reminders of this remarkable naturalist.

Franklin's Gull

Franklin's gull is every bit as graceful a bird as Forster's tern, but being a gull it has broader, less pointed wings as well as a more robust bill suited to its insectivorous diet. These are smallish, black-headed gulls that often wheel in flocks over western plains, frequently following plows to pick up insects. Early settlers called them "prairie pigeons." In the summer they settle in large colonies to nest on lakes and marshes from Utah and Colorado into central Canada. In the fall they migrate to the Gulf coast or on into Central and South America. It was in Saskatchewan that members of the expedition of John Franklin (1786–1847), passing through Canada to the Arctic coast, took specimens that were later named for Franklin. Franklin's spruce grouse and Franklin's ground squirrel are other products of the expedition. Franklin was more an explorer than a naturalist; his companion John Richardson did most of the collecting.

As a young man, Franklin sailed around Australia with his uncle, Matthew Flinders, and he later fought in the battles of Trafalgar and New Orleans. He had already participated in an unsuccessful attempt to reach the North Pole when he was selected to lead an expedition to explore the north coast of Canada as part of the search for a sea passage across the north of the American continent. He and his men arrived at York Factory, on Hudson's Bay, in August 1819. It took Franklin and his men two winters and another summer to establish a base camp near Great Slave Lake, from which, in the summer of 1821, they reached and explored part of the Arctic coast. On their return trip they met deep, early snows and had to abandon most of their supplies and equipment. For a time they lived on lichens, old shoes, and a boiled buffalo robe. Several men did not survive, and one member of the expedition was suspected of murder and cannibalism. Franklin and other survivors arrived back at their base camp with "ghastly countenances, dilated eyeballs, and sepulchral voices."

Franklin made it back to England only to return to the Arctic coast by a similar route three years later, this time with both Richardson and Drummond as naturalists. This was a better planned expedition and much more productive in terms of natural history. Many of the results were included in W. J. Hooker's *Flora Boreali-Americana* and Richardson's *Fauna Boreali-Americana*, the latter with a section on birds by William Swainson, who named Franklin's gull. Both Franklin and Richardson were knighted.

In 1836 Franklin was posted to a seven-year stint as governor of Tasmania. He seemed ill-suited to governing this tumultuous colony, which had been receiving great numbers of convicts, although his wife Jane did much to improve the social life of Tasmania. Back in England, Franklin was asked to undertake still another effort to find the Northwest Passage, although he was nearly sixty years old. In 1845 he sailed for the Arctic with two ships, the *Erebus* and the *Terror;* there were 134 officers and men and supplies for three years. Richardson did not accompany Franklin, but a few years later he made an expedition of his own—in a vain attempt to rescue Franklin, who never returned. It was eventually learned that his ships had been icebound for many months. Franklin died on the *Erebus* before the ships had to be abandoned; the remaining members of the expedition struggled south by land and died one by one without finding help. Franklin was said to be a kindly man, much admired by his subordinates, but he must have had a spirit of steel to survive so many adventures before finally yielding to odds too great for any man.

Fremontia *(Flannel Bush)*

John Charles Frémont (1813–1890) was also primarily an explorer who, like Franklin, was greatly admired in his time. As a young man in South Carolina, Frémont attracted the attention of Joel R. Poinsett, who had just come back from his post as ambassador to Mexico with the plant named for him, now so closely associated with Christmas: *Poinsettia*. Frémont was appointed a lieutenant in the U.S. Topographic Corps and assigned to expeditions to the Carolina mountains and to the Missouri River. Back in Washington, he met the powerful senator Thomas Hart Benton and shortly eloped with his sixteen-year-old daughter Jesse. The young Frémont was handsome, brilliant, ambitious—and lucky, with supporters such as Poinsett and Benton, and Jesse, who stuck with him throughout a turbulent career.

Frémont's three expeditions through the West as an officer of the Topographic Corps were all productive botanically, even though he declined to take an experienced botanist with him. "Frémont [wrote George Engelmann] appears to me rather selfish, disinclined to let anybody share in his discoveries, anxious to reap all the honors as well as undertake all the labor himself." His reports, ably edited by Jesse, contain many references to plants by their scientific names, but these were, for the most part, added by John Torrey as the journals were prepared for publication. Torrey stated that on the second expedition about fourteen hundred specimens were collected, "many of them in regions not before explored by any botanist.... [However] more than half of his specimens were ruined before he reached the borders of civilization." On one occasion a mule laden with specimens plunged over a cliff and could not be rescued; on another occasion all were lost in a flash flood.

Nevertheless Torrey, as well as Gray and others, described many plants from Frémont's expeditions, among them *Fremontia californica*, or flannel bush, a southwestern shrub that bears showy masses of yellow flowers. Other plants bearing Frémont's name include Frémont's phacelia, Frémont's monkey flower, and Frémont's geranium. Four subspecies of mammals are also named for him. *Tamiasciurus hudsonicus fremonti* was named by Audubon and Bachman; usually this subspecies of the red squirrel is called the pine squirrel, but Colorado naturalist Enos Mills preferred "Frémont's squirrel" when he wrote about it in his essay "A Midget in Fur." Frémont Peak stands 13,730 feet in the Wind River Range of Wyoming; when Frémont climbed it in 1842 he believed it to be the highest in the Rockies.

On his third western trip Frémont became involved in the conquest of California but fell afoul of General Stephen Kearny. He was court-martialed and found guilty of insubordination, but President Polk remitted the sentence. Frémont left the army, and as a private citizen, with Benton's blessing, he undertook a fourth western expedition. He tried to cross Colorado's San Juan Mountains in the winter, but eleven members of his party died in the attempt, and there were rumors of cannibalism. Despite this disaster, Frémont was nominated for the presidency in 1856. He did not wage an effective campaign and was defeated by James Buchanan. He served in the Civil War, though without distinction. After the war he speculated in mining and railroads, but he met with little success and died in New York in relative poverty. Thus his life of adventure, like Franklin's, ended sadly, even more so because he died far from the scene of his earlier triumphs and after a long decline in influence and popularity.

Fendlera *(False Mock Orange)*

The name of August Fendler (1813–1883) is scarcely as well known as that of Franklin or Frémont, but he, too, lived an adventurous life. Arriving from Germany in 1836, Fendler wandered about the West for a time, doing some teaching, working in a refinery, and collecting a few plants, which he sent to George Engelmann in St. Louis. When Asa Gray learned that an army expedition was being sent to Santa Fe, just recently taken from the Mexicans, he asked Engelmann if he knew of someone who could go along as a naturalist. Engelmann suggested Fendler, and the choice proved to be a good one. In a letter to Engelmann, Gray remarked: "I got a parcel [from Fendler]. The specimens are perfectly charming! so well made, so full and perfect. Better were never made." Gray's publication *Plantae Fendleriana* included descriptions of the many novelties Fendler had discovered. Fendler was, of course, paid for his collections, though he sometimes complained that he received hardly enough to keep from starving. In 1849 Fendler set out on a trip to the Great Salt Lake, but on the way he lost nearly all his supplies in a flash flood. Back in St. Louis, he found that everything he owned had been lost in a great fire. He later went to Venezuela, still collecting for Gray. He died in Trinidad at the age of seventy.

Since Fendler was among the first to collect plants in New Mexico, it is not surprising that many are named for him. *Fendlera rupicola*, called false mock orange or fendlerbush, is a shrub up to six feet tall, bearing showy white flowers and later small, acornlike fruits. It is a favorite browse of goats, deer, and bighorn sheep. *Fendlerella utahensis* has small, white blossoms in clusters; like *Fendlera* it is a member of the saxifrage family and occurs in rocky places in the Southwest. *Ceanothus fendleri* is still another attractive shrub, bearing clusters of small, white, sweet-smelling flowers. Other plants named for Fendler include Fendler's aster, Fendler's rock-cress, and several grasses, mosses, and lichens.

Forsythia

Finally, a word should be said for William Forsyth (1737–1804), even though he lived a quiet life in London as supervisor of Kensington Gardens and never traveled to the New World. He was the author of books on sylviculture and a member of the Linnaean Society of London. The plant named for him, *Forsythia*, has become a popular ornamental, its abundant yellow flowers appearing in early spring before the leaves unfold.

Gambel's Quail

There is something appealing about quails. Running about in small coveys through the brush, they seem to chatter among themselves as they pick up seeds and plant fragments. Often they appear relatively tame, but when frightened they scatter on their short wings, calling to one another in alarm. After the danger passes they reassemble and once again call quietly. In season, they build a simple nest on the ground, where the female lays as many as a dozen eggs. When the chicks are born they follow their mother for a few days, although they are able to feed themselves almost immediately, preferring insects, which provide them with the protein they need for growth.

Gambel's quail (*Callipepla gambelii*) is a bird of the deep Southwest, living in places where creosote bush, mesquite, and cacti are the dominant plants. However, they do require water, as the young chicks must drink within a few hours after they hatch. Gambel's is perhaps the handsomest of the quails. Adult males are grayish above, with a red cap and a black throat outlined by white; beneath, they are buff, with a central black spot and streaked, chestnut sides. The most striking feature is a group of feathers on top of the head, about one and a half

Gambel's Quail

inches long, and curving forward, rather like the plume in the hat of a medieval cavalier.

What can this ornamentation mean in a modest little bird that spends its time puttering about in the deserts? Perhaps the birds need this elaborate plumage so that they can recognize members of their own species and avoid mating with members of alien species. But this seems unlikely for two reasons: the females of this and related species are not nearly as fancily clad, and in fact hybrids have occasionally been found. A better explanation is to be found in concepts of sexual selection, as promulgated first by Charles Darwin. Males compete for mates, and females choose to mate with superior males, males that will ensure that their offspring are superior. A superior male is one that vigorously displays the excellence of his plumage (and probably is more vigorous in every way). Thus, over many generations, there is "runaway" selection for the elaboration of ever more bright colors, fancy plumes, and complex courtship displays. But these features may make males more conspicuous to predators (in fact, they often do). So in time a balance is struck; males are as brilliantly colored as they can afford to be in the context of their life-style and natural enemies. If males live in deep forests and do nothing to help rear their young, sexual selection may lead to fantastic results, as in male birds of paradise. Gambel's quails have by no means gone that far; they are far too exposed to marauding coyotes and to hawks and eagles from above.

Like many other inhabitants of the desert, Gambel's quails depend upon winter rains to produce an abundant green growth in spring. This new growth provides the foodstuffs needed for the development of their reproductive organs and assures a good supply of seeds and insects for the chicks. When the winter rains fail, the reproductive organs remain dormant and mating and egg-laying are postponed until a more favorable season.

It was on his return from a trip through the West, in 1843, that William Gambel (1821–1849), then only twenty-two, submitted to America's most prestigious scientific establishment, the Academy of Natural Sciences of Philadelphia, a description of several "new and rare birds" that he had discovered. One of these was Gambel's quail, a "handsome species . . . inhabiting the most barren brushy plains . . . where a person would suppose it to be impossible for any animal to subsist." The birds occurred in "small flocks of five or six, occasionally uttering a low gutteral call of recognition. . . . When flying they utter a loud sharp whistle, and conspicuously display the long crest."

Gambel named the quail *gambelii* and ascribed the name to Thomas Nuttall, who had presumably told Gambel that he had described it earlier and named it for him. However, Nuttall had evidently not gotten around to it, so Gambel's description is the first in print. Thus Gambel appears to have named the species after himself, something that is *never* done. But in this case it is forgivable.

As a youth, Gambel had worked as Nuttall's assistant, and it was doubtless Nuttall who encouraged him, a lad of twenty, to join a group of trappers who were headed down the Santa Fe Trail. At Santa Fe he joined another group following the Mormon Trail to California. Along the way, and for several months in California, he collected plants and animals, many of them new to science. Back in Philadelphia, he served as assistant curator at the Academy of Natural Sciences for several years. Then in 1849 he joined a second expedition to the West, this under the command of General Isaac Wistar, who described Gambel as "a genial, kindly man and delightful companion, but averse to a rough life [and] hard work." Doubtless so that he would have more time to collect specimens, Gambel left to join a party led by "Captain Boone," one of Daniel's numerous descendants. Boone's party met with misadventures while crossing what is now Nevada and was forced to cross the Sierra in winter. Most of the group perished, and two survivors later gave Wistar a report on their fate. Wistar wrote to a friend in part as follows.

> After the loss of most of their cattle and consequent abandonment of many wagons in the Humboldt Desert, they were caught by snow in the mountains, and instead of abandoning the remainder and pushing through, they camped to await better weather, which did not come. But few got across the range, including Gambel, and these saved little but what they stood in. With numbers rapidly diminishing the remnant pushed down to Rose's Bar [on the Feather River of California], where several, including Gambel, died almost immediately of typhoid fever. Gambel was buried on the Bar which, however, has since been entirely removed by hydraulic mining.

Gambel was only twenty-eight when he died. So valuable and well prepared were his specimens that several plants and animals were named for him, including not only the quail but one of the southern Rockies' most characteristic trees, Gambel's oak (*Quercus gambelii*), named by Nuttall. Spencer Baird named the genus of the leopard lizards *Gambelia*, and he and Charles Girard named a species of these lizards *Gambelia wislizenii*, thus causing a rather small reptile to carry within him the souls of two notable naturalists. (Frederick Adolphus Wislizenus was a protégé of George Engelmann's who traveled the western states and Mexico in the 1840s.) Gambel himself described several important western birds, including Nuttall's woodpecker, the California thrasher, the wrentit, and the elegant tern.

Gambel also described the mountain chickadee, calling it *Parus montanus*. However, this name had been used previously for another species, so the mountain chickadee had to be renamed to avoid confusion. This Robert Ridgway did in 1886, calling the species *Parus gambeli*. Ridgway (1850–1929) came from a Quaker family that had settled in Illinois. He acquired a love of nature from his parents and as a youth began studying the birds around his home. He wrote to Spencer Baird concerning some birds that puzzled him, and this led to a long correspondence and, some years later, to his appointment as curator of birds at the U.S. National Museum. In the meantime, when he was only seventeen, he received an appointment as zoologist of the Geological Exploration of the 40th Parallel, commanded by Clarence King. The expedition ran into many problems with weather and illnesses in the Far West of the 1860s, some of them in the same Humboldt Sink that had delayed Gambel and contributed to his death. Ridgway's report on the ornithology of the expedition made an excellent impression and led to his government appointment, which he held for many years.

Ridgway was one of the founders of the American Ornithologists' Union in 1883 and one of its early presidents. He was especially concerned that naturalists describe the colors of their specimens accurately, and to this end he published his *Color Standards and Nomenclature*, in which various shades of colors were given precise names. After publishing voluminously on birds, he retired to Illinois and established a wildlife sanctuary which he called Bird Haven. It has been kept as a memorial to him. Ridgway had an aversion to public speaking, but he was a skilled writer of both scientific and popular articles and books on natural history themes. His one son, Audubon, followed in his footsteps as an ornithologist but unfortunately died when he was only twenty-four. Ridgway never had a genus or species named for him—such was the fate of those who came along after most of the pioneering work in their field had been completed. He did have at least four subspecies of birds named for him.

Gregg *Dalea*

When William Gambel traveled down the Santa Fe Trail in 1841, he was following in the tracks of Josiah Gregg (1806–1850), who had accompanied a caravan down the trail ten years earlier, hoping to recover his health in the dry atmosphere of the Southwest. His health improved, and he adopted the life of a trader, making frequent trips from Independence, Missouri, to Santa Fe and sometimes on to Chihuahua. Gregg was well educated, and he was a close observer of nature as well as of the motley groups of people that traversed the trail. In 1844 he published *Commerce of the Prairies*, a book that went through six editions in English and three in German up to 1857. The book became a classic of western history and is still in print in paperback. After spending some time in Mexico, Gregg set out in 1850 with several others to explore parts of California. The group suffered from dissension within the ranks and from hunger and exposure. According to one report, Gregg died of starvation; according to another, he fell from his horse when in a weakened condition and died shortly thereafter. He was only forty-four when he died.

Commerce of the Prairies contains few natural history observations, but Gregg was nevertheless collecting specimens from time to time and sending plants to George Engelmann, who forwarded many of them to Asa Gray. Gray was generous with eponyms, naming *Greggia*, a genus of cruciferous plants from Mexico, *Acacia greggii*, catclaw acacia, *Fraxinus greggii*, an ash, and *Dalea greggii*, Gregg dalea. The last is a handsome shrub with silvery leaves and rose-purple flowers, adorning rocky slopes in the deep Southwest and sometimes used for xeriscaping. To these tributes Engelmann added the beautiful night-blooming cactus *Peniocereus greggii*.

Gunnison's Prairie Dog

Like Josiah Gregg, John W. Gunnison (1812–1853) was only marginally a naturalist. He was trained at West Point, and as an officer in the U.S. Topographic Corps he was involved in surveys in several parts of the country before being assigned to explore a central route for a railroad to the Pacific. This took him through central Colorado to Salt Lake City, where he was snowed in for several months and occupied his time by writing a book on the Mormons. On a second trip into Utah, in 1853, he was accompanied by a botanist from St. Louis, Frederick Creutzfeldt, who had barely survived Frémont's ill-fated fourth expedition. This time he did not survive, nor did Gunnison, for most of the exploring party was slaughtered and mutilated by Paiute Indians on the Sevier River. The Paiutes were evidently seeking revenge for an attack by Whites a few weeks before.

The city of Gunnison, Colorado, and the Gunnison River, with its Black Canyon, now a national monument, all serve to remind us of the accomplishments of this hardy explorer. And two very special animals and plants bear his name: Gunnison's prairie dog, *Cynomys gunnisoni*, and the elegant sego lily, *Calochortus gunnisonii*. The bulbs of sego lilies saved many an early settler from starvation; the Mormons, in particular, consumed many during their first difficult years in Utah. Gunnison's prairie dog occurs in mountain valleys in the southwestern states, even up to the timberline. Unlike other prairie dogs, it does not make large mounds at burrow entrances and does not form dense colonies, so it is not disliked quite as fervently as ranchers dislike black-tailed and white-tailed prairie dogs.

Gambel, Gregg, and Gunnison died within four years of one another, in the prime of their lives and while seeking to bring a little more of the unknown into the sphere of human knowledge. Such were often the lives of frontier naturalists.

Henslow's Sparrow

*H*enslow's sparrow (*Ammodramus henslowii*) is well known only to ornithologists and to the most dedicated of bird-watchers. In the summer it ranges throughout the northeastern quarter of the United States, but it is confined to damp, weedy fields and meadows, where it is the most secretive of birds. In the spring, the males establish territories about two acres in size and announce their territorial rights with a rather weak "tsi-lick" from the top of a tall weed—in Roger Tory Peterson's words, "one of the poorest vocal efforts of any bird. . . . As if to practice this 'song' so that it might not always remain at the bottom of the list," he adds, "it often hiccoughs all night long." The female remains close to the ground, stealing mouselike among the grasses. The nest is built of grasses close to the ground, where three or four eggs are laid in June. The nestlings are fed grasshoppers and other insects by both parents. By September all are ready for the short flight to the southern states, where they spend the winter in fields and pine woods.

It was Audubon who discovered the birds in Kentucky, across the river from Cincinnati, in 1820. Then thirty-five, he was working as an artist and curator in Dr. Daniel Drake's Western Museum. Nine years later he included a painting of the sparrow

in his *Birds of America* against a background of Indian-pink, as
if to add color to the portrait of a relatively drab bird. Audubon
faithfully showed the thick bill and the chestnut coloration of
the back and wings. This is one of the smaller sparrows, a good
deal smaller than a song sparrow, with a short tail and a streaked
breast, separable with some difficulty from other "little brown
birds" of fields and meadows.

Following Audubon's discovery, these birds were not
recorded from Kentucky for another sixty-five years. Such is the
history of the species—sometimes locally common but often
disappearing from a site for several years. Apparently its dependence upon fields of a particular nature explains its spotty
occurrence. Their ideal habitat is a field not recently burned or
heavily grazed, with tall grass, rank weeds, and a few bushes.
Such fields often do not endure for long; they tend to be put to
use for agriculture or grazing, or if left alone to grow up with
brush and trees. Although Henslow's sparrow has never been

Henslow's Sparrow

common, except locally and occasionally, it is apparent that populations have been declining in recent years.

In this respect Henslow's sparrow is following the pattern of many songbird species, for there is now firm evidence that many North American and Eurasian birds are declining in numbers. Rachel Carson, in *Silent Spring*, blamed DDT and other pesticides. She was surely on the right track, but the decline continues even though DDT and several other pesticides have been banned in the United States. (They are still manufactured here and sold to countries where many of our birds overwinter.) The fact that birds that overwinter within the United States (like Henslow's sparrow) are declining suggests that loss of habitat is of overriding importance, as it surely is also in the tropics. This is an unpreventable disaster so long as humans continue to increase in number and to insist on more "efficient" use of the landscape. Perhaps Henslow's sparrows will be missed by few, though every species, as the product of eons of evolution, is surely very precious. Other birds that are in serious decline, such as the wood thrush, will be missed by all sensate persons.

Audubon named the species for a person who had never visited North America but whom he had come to respect and feel grateful to. When Audubon met him, John Stevens Henslow (1796–1861) was a professor of botany at Cambridge University as well as curate of St. Mary's church in Cambridge. Audubon was in England seeking subscriptions for his *Birds of America* and looking for advice on booksellers. His reputation had preceded him, and even King George IV took time from his whist games to look at his paintings. Other notables invited him to their homes, including Professor Henslow, who became an enthusiastic subscriber.

Charles Darwin was then a student at Cambridge, and though he was bored by most of his course work, he liked Henslow's lectures "for their extreme clearness, and the admirable illustrations." As Darwin wrote in his *Autobiography*: "Henslow used to take his pupils, including several of the older members of the university, [on] field excursions, on foot or in coaches, to distant places, or in a barge down the river, and lectured on the rarer plants and animals which were observed. These excursions were delightful."

Darwin spent many evenings with Henslow and took long walks with him. On returning home from a trip to Wales in 1831, Darwin found a letter from Henslow telling him that Captain Fitzroy, of the *Beagle*, had offered to share his cabin for a trip around the world with a young naturalist who would come without pay. Henslow recommended Darwin, who of course accepted, and it was to Henslow that Darwin shipped back cases of specimens from diverse parts of the world. He wrote to Henslow as often as he could, and Henslow often read his letters before various scientific societies. The novel ideas that Darwin presented after his return were, however, a little hard for a devout cleric to take. Darwin wrote to Asa Gray that "Henslow will go a very little way with me and is not shocked by me." At the meeting of the British Association for the Advancement of Science in 1860, at which "Darwin's bulldog," Thomas Henry Huxley, put down Bishop "Soapy Sam" Wilberforce, it was Henslow who presided. He died only a few months later, perhaps never fully convinced though surely pleased at the impact his former student had made. Henslow deeded his extensive collections of plants to Kew Gardens, which he had helped to found.

Audubon never met Darwin, and his contacts with Henslow were brief, though he was obviously impressed by him. It hardly seems necessary to say much about John James Audubon (1785–1851), who has been the subject of several biographies and whose name is joined with the names of many of the naturalists profiled in these pages. He was born out of wedlock in Haiti but adopted by his father and taken to France for his education. His French accent remained with him all his life, adding to the romantic aura that surrounded him as he later roamed America's wilds with his flute, his violin, his rifle, and his paints. Audubon began drawing birds even before he moved to America, at the age of eighteen, and once settled on his father's farm in Pennsylvania, he began to collect and paint birds and mammals intensively. He also courted and won the heart of a neighborhood girl who had recently come from England, Lucy Bakewell. Lucy is said to have acquired a love of nature from the family physician, Erasmus Darwin, grandfather of Charles.

After failing in several business ventures, Audubon, at thirty-four, began to expand his portfolio of bird paintings with the idea of publishing them. This led him to search for more birds to paint, chiefly in the Ohio and Mississippi valleys. Lucy was often left to support herself and the children by serving as governess or companion. With only his dog with him, Audubon followed trails through the forests, usually stopping at the cabin of a settler in the evening, where he was often treated to a simple meal and regaled with stories. He went hunting with Daniel Boone, then an old man but still a crack shot. Each day he wrote in his journal, describing not only the wildlife he had observed but also the lives of the people of the frontier. Many of his adventures would eventually be incorporated into his *Ornithological Biography*.

From time to time Audubon earned a few dollars by teaching or by painting portraits. When he visited Philadelphia and New York in 1824, he found no one able to help with the publication of his portfolio of paintings. With hard work, he and Lucy managed to put together enough money to send him to England, where the "American woodsman," still with his long hair and frontier garb, entertained his newfound friends with tales of the wilderness. Here he found an engraver for his paintings and many subscribers for his *Birds of America*. In Paris, the great zoologist Georges Cuvier spoke of Audubon's paintings as "the greatest monument ever erected by art to nature."

After a second trip to England, this time with Lucy and his son John, Audubon found that he needed to collect more birds to round out his volume, so he traveled to Florida, to Labrador, and to the upper Missouri basin, where he hunted bison. The ultimate success of *Birds of America* led him, late in life, to undertake a similar project on the mammals of North America, done in collaboration with his friend John Bachman and with the help of his two sons, who had married Bachman's two daughters.

Audubon became acquainted with most of the naturalists of his day, and he also met with presidents, kings, and other people of influence. He was an eponym addict, and the species he named honor many of those he admired or shared experiences with. Bachman's warbler was, of course, named for his good friend and collaborator. Baird's sparrow was named for Spencer Baird, whom he had encouraged as a youth and who went on to become secretary of the Smithsonian Institution. Lincoln's sparrow he named for Thomas Lincoln, who accompanied him on his trip to Labrador, Sprague's pipit for Isaac Sprague, a companion on his trip up the Missouri. Bewick's wren was named for Thomas Bewick, whose engravings of British birds Audubon admired, and Bell's vireo for John Bell, a well-known New York taxidermist.

There were others. But Audubon himself fared poorly. Audubon's oriole is a handsome enough bird that is largely Mexican in distribution. Audubon's warbler was found to be no more than a color form of the yellow-rumped warbler, so the name is no longer used. The name Audubon's shearwater is accepted by the American Ornithologists' Union on the basis of Audubon's painting. However, the species was earlier named *Puffinus lherminieri* by French ornithologist René Primevère Lesson, honoring a father and son team of Guadaloupe naturalists. The desert cottontail rabbit was named by Baird *Sylvilagus auduboni*, a tribute to Audubon's contributions to the study of mammals. Of course, we are not likely to forget Audubon's name; the national and regional Audubon Societies are named for him, and Mount Audubon stands 13,223 feet high in the Rockies, not far northwest of Denver.

Audubon's success owes much to his long-suffering wife Lucy. As mentioned earlier, Spencer Baird named his daughter after Lucy Audubon, and James Cooper named Lucy's warbler after Lucy Baird. So in a sense Lucy Audubon has a hand-me-down eponym.

Harris's Hawk

Among other birds named by Audubon for special friends were two impressive raptors, Harris's hawk and Harlan's hawk. Harris's hawk is a denizen of the deep Southwest, a handsome bird with chestnut shoulders and leggings and a white tail-band against an otherwise dark brown plumage. Sometimes these hawks can be seen atop a saguaro cactus, scanning the desert for jackrabbits. These are most unusual hawks, since they often hunt in small groups. When a rabbit is spotted, one may flush it from a hiding place, while a second pounces on it; sometimes three or four give the rabbit a merry chase, then share the kill or take it to the nest to feed the young. These cooperating individuals are members of a family, which includes one female and two or more males. In contrast, most species of hawks are solitary hunters. John Alcock has written a vivid portrait of Harris's hawk in his book *Sonoran Desert Summer*.

Edward Harris (1799–1863) was a gentleman farmer of Moorestown, New Jersey, who befriended Audubon when he needed financial help during the preparation of his *Birds of*

America. He accompanied Audubon on his trip to Texas in 1837 and on his trip up the Missouri River in 1843. Audubon described him as "a gentleman to whom I am most deeply indebted for many acts of kindness and generosity, and in particular for his efficient aid at a time when, like my predecessor Wilson, I was reduced to the lowest degree of indigence.... But, independently . . . he merits this tribute as an ardent and successful cultivator of ornithology. . . ." Audubon also named an unusual sparrow for Harris. Harris's sparrow is larger than most of its kin and bears a black cap and bib that makes it easy to recognize. These birds breed in the Far North but are sometimes seen migrating in flocks across the Great Plains.

Harlan's Hawk

Harlan's hawk, which also breeds in the Far North, is now considered no more than a dark subspecies of the common redtailed hawk. Richard Harlan (1796–1843) was a Philadelphia surgeon with an interest in zoology and paleontology. In 1825 he made a field trip with Thomas Say and Titian Peale, resulting in the discovery of the first American plesiosaur. A few years later he published a paper on *Megalonyx*, an odd fossil animal that Thomas Jefferson had puzzled over. Harlan foresaw the time when the remains of great dinosaurs would be found in the West. His major book, *Fauna Americana*, was the first systematic treatment of all known American mammals, living and fossil, but it was clearly the work of an amateur and was not well received by many naturalists. Audubon, however, admired him and sought a bird to name for him "which in size and importance should bear some proportion to my gratitude toward that learned and accomplished friend."

Holbrookia *(Earless Lizards)*

John Edwards Holbrook (1794–1871) was also a friend of
Audubon, who met him during one of his visits to his friend
and collaborator John Bachman in Charleston, South Carolina,
where Bachman was a prominent clergyman-naturalist and
Holbrook a leading physician-naturalist. Holbrook had received
his medical degree at the University of Pennsylvania and had
studied in Europe for several years before settling in
Charleston. He helped to found the Medical College of South
Carolina and was a professor there for over thirty years. While
he was in Europe, French naturalists encouraged him to undertake a study of North American reptiles and amphibians, and he
set about to do so. Holbrook's monograph *North American Herpetology* was the first in its field and was illustrated with many
color plates. Audubon had sent him a rattlesnake from Texas for
inclusion in his book. Later Holbrook prepared a similar monograph on the fishes of South Carolina. When Harvard guru
Louis Agassiz lectured in Charleston in 1848, he was entertained by Holbrook on his plantation, where Agassiz walked
through the fields observing the slaves. Slavery, he mused,
"might some day be the cause of the ruin of the United States."
The Civil War was surely the ruin of Holbrook's career. He
served as a medical officer in the Confederate army and after
the war lost all of his books and collections.

Holbrook had nevertheless made his mark as America's pioneer herpetologist. Agassiz's former student Charles Girard
named the genus of the earless lizards *Holbrookia*. These are
small lizards that occur in sandy soil in the western states, feeding on insects and spiders. They appear to have a curiosity
unexpected for so small-brained a creature. Workers in a quarry
once saw one take up an observation post nearly every day,

"placing his feet on a clod so as to be sure not to miss anything."
Several species of *Holbrookia* are known—but none of them
very well known, as these are not glamorous animals and have
few aficionados.

Besides the lizards, Holbrook's king snake and Holbrook's
spadefoot toad are named for him (the latter by Richard Harlan). Holbrook himself named many species and did not always
confine himself to describing dead museum specimens. He
wrote, for example, of the bell-like quality of the song of the
green tree frog and how when one frog began to sing, "hundreds
take it up from all parts of the corn field, and when he stops, the
concert is at an end, until he again begins." It is in the field,
after all, where naturalists are most at home; preserved specimens simply make the winters more bearable.

Ibsen's Wasp

*I*bsen's wasp (*Pardiaulomella ibseni*) is 1.8 millimeters long—about one-sixteenth of an inch—just the right size to perch comfortably on the head of a pin. So it is not likely to strike fear in even the most committed wasp-hater. Under a good microscope, it is a rather beautiful creature, metallic blue with white legs and bright red eyes. Entomologists call the group to which it belongs chalcid wasps, based on the Greek word *chalkos*, copper, with reference to the metallic coloration of most of them. Despite their diminutive size, these wasps play important roles in nature, since there are many thousands of species, many of them parasitic on noxious insects. Ibsen's wasp was discovered in 1916 by a researcher of the U.S. Department of Agriculture, J. F. Strauss, who was studying the life history and control of the grape leaf-folder, a pest of grapes throughout much of the United States. The yellow-green larvae of the pest fold over the leaves of grapes, tie them with silk, and proceed to skeletonize the leaves. The moth that is eventually produced has a wingspan of about an inch; the wings are black, with two white spots on each front wing and one on each hind wing.

Strauss reared several parasites from the caterpillars, including one chalcid wasp that was new to science. He submitted

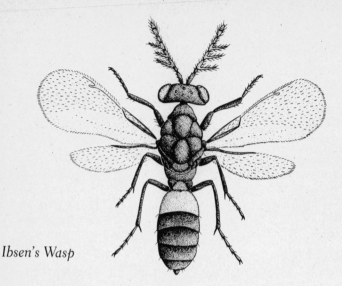

Ibsen's Wasp

specimens to taxonomists of the Department of Agriculture, where they found their way to a specialist, Alexandre Arsène Girault (1884–1941). Without explanation, Girault named the species after Norwegian playwright Henrik Ibsen (1828–1906), author of *Peer Gynt, Hedda Gabler, An Enemy of the People,* and other noteworthy plays. It may be that Girault had a special affection for Ibsen, who has been described as a "protagonist of the individualists and the heroic minority." Girault was most certainly an individualist, so much so that he continually quarreled with his colleagues and his superiors. Within a few months after he described and named Ibsen's wasp, he was discharged by the Department of Agriculture and moved to Australia (where he had been for three years previously). But here things were no better, and he moved from one job to another, even for a time running a poultry farm, another time a fruit stand. Eventually he returned to professional work, but this terminated in 1935, leaving him with five children to support (his

wife had died a few years earlier). At one time he even worked
in a stone quarry, his "final humiliation." But up until his death
he was devoted to chalcid wasps. "I have wooed, won and
enjoyed these worthy woodland spirits for their and Love's
sweet sakes [he wrote in 1923]. I . . . name them that others
may know what Grace and Beauty the woodland holds."

Needless to say, with such an unconventional writing style he
was not beloved of the editors of scientific journals, and he soon
parted company with them and began to publish his papers privately. Altogether he put out sixty-three publications at his own
expense, in one specialist's words "a rare mixture of science,
libel and other matters of dubious taste." He was not a bad ver-
sifier. Here is his tribute to W. H. Ashmead, one of his prede-
cessors as a chalcid expert at the U.S. National Museum:

> *False Captain! Ah! dark Error's pioneer,*
> *Enthusiastic dunce and shamming seer,*
> *Aching for a day's applause;*
> *Low scholar ever wishing us to laud*
> *Ambition's wind-blown froth and sandy fraud,*
> *Thus defying Heaven's laws.*
>
> *Arise! Come, get thee from thy shelt'ring grave*
> *Where, strongly walled, e'en thou couldst dare be brave*
> *With Impunity's gaunt grace!*
> *Ah, come, past coward, lily-livered liar,*
> *Fair-tongued sweetmouthing unctious friar*
> *Let's see what's writ across thy face!*

In the midst of snatches of verse, philosophy, and vitupera-
tion aimed at colleagues and superiors, Girault managed to
describe many hundreds of chalcid wasps. Doubtless at times
he was hard put to find names for them, and his naming of a
species for Ibsen was not uncharacteristic. He also named

species after Shakespeare, Goethe, Carlyle, Voltaire, Cervantes, Longfellow, Wordsworth, Titian, Botticelli, Holbein, and other figures from the world of art and literature. He did not neglect composers: there were species named after Chopin, Schumann, and Haydn, as well as one called *Mozartella beethoveni*. He even waxed political at times, naming species after Garibaldi and Lenin. Despite his overly brief descriptions, his rough treatment of specimens, and the fact that many of his papers were published privately, his names have been reluctantly accepted. Several species have even been named for him, as well as one genus, *Giraultia*.

Girault may have been the prototypical "crazy entomologist," but he is secretly admired by some of us who allow ourselves to cater to the conventions of society. Alas, I shall probably never meet with Ibsen's wasp and surely not with *Mozartella beethoveni*, which hails from Pinkenba, Australia. But I hope they are thriving, as well as others of Girault's "worthy woodland spirits." Girault himself will haunt the history of entomology for many years to come.

Some Entomological Eponyms

Entomologists are particularly fond of naming insects for their colleagues or for persons they admire. After all, there are perhaps a million species of insects, and entomologists are sometimes desperate for names. One acquaintance of mine found himself confronted with more than a dozen previously unknown species requiring names. He named one for his wife, others for his son, daughter, mother, father, grandmother, nephew, and so forth—even the mother of his sister-in-law. Another acquaintance named a species for his wife; she later divorced him, but she had no way of shedding the "bug" named for her. Even W. H. Ashmead, so maligned by Girault in the verse quoted above, had an attractive genus of bees named for him, *Ashmeadiella*. Indeed, it is hard to think of a prominent systematic entomologist who has not had a genus or a few species named for him or her.

Naturalists mean to honor the persons for whom they name plants or animals, but sometimes the results are questionable. The genus *Dyaria*, for example, was named for a distinguished student of moths and of mosquitoes, Harrison G. Dyar, without appreciation of the evocation that the generic name's pronunciation would likely have. A genus meant to honor Henry Townes, a world authority on ichneumon wasps, was named *Townesilitus*.

Generally speaking, entomologists are remarkably tolerant of eponyms that others might consider insulting. John L. LeConte had a beetle named for him, *Philonthus lecontei* (*Philonthus* is based on the Greek words *philia*, affection, and *onthos*, dung). LeConte doubtless felt honored; beetles, of all kinds, were his life. And certainly Thomas Nuttall had no objection when his friend Thomas Say named a blister beetle *Lytta nuttalli* (*lytta* is Greek for "madness"; when Nuttall traveled up the Missouri, French voyageurs called him "le fou" for his singleminded devotion to collecting specimens, but it is doubtful Say had that in mind).

There is less to be said for the invention of frivolous names. There is, for example, a beetle named *Ytu brutus* and a wasp named *Aha evansi*, the latter based on specimens I collected in Australia. The British entomologist G. W. Kirkaldy carried things to an extreme when he named a genus *Ochisme* (pronounced "oh-kiss-me") and followed it with *Polychisme*, *Dolichisme*, and *Peggichisme*. The detailed delineation of attributes required in writing up the description of a new genus or species can be monotonous, so perhaps occasional attempts at humor are excusable.

Jamesia

*J*amesia, also called waxflower, cliffbush, mountain mock orange, or wild hydrangea, is a handsome shrub that grows in rocky ledges from southern Wyoming to New Mexico and west to California. The waxy white flowers are faintly fragrant and are born in clusters at the tips of branches three or four feet tall. The leaves are somewhat downy and have jagged edges; in the fall they turn rusty red. Westerners are inordinately fond of these plants, which do so much to brighten rocky outcrops and cliffsides. Jamesia is the only western representative of the hydrangea family and one of only a few North American members of this family. For a long time it was thought that there was only one living species, *Jamesia americana*, but researchers at the New York Botanical Garden have recently found a second species in Nevada and western Utah. A fossil species of *Jamesia* has, however, been found near the old mining town of Creede, in southwestern Colorado, in rocks twenty-six million years old. So evidently these are ancient plants that have survived by clinging to a habitat where few other plants do well. The fossils are well preserved in fine-grained shales that were formed when sediments were deposited in a caldera lake.

The shrubs were discovered by Edwin James (1797–1861), a member of Major Stephen Long's expedition of 1820 to the Rocky Mountains. This was a small, ill-equipped expedition that moved rapidly down the eastern edge of the Front Range, from near the present site of Denver to near Pueblo. No one is sure where James collected the specimens that he later sent to John Torrey and Asa Gray, leading botanists of the time. Torrey and Gray described *Jamesia americana* in 1840, with the following dedication: "[We] have applied the present name in commemoration of the scientific services of its worthy discoverer, the botanist and historian of 'Major Long's Expedition to the Rocky Mountains, in the year 1820,' and who, during that journey, made an excellent collection of plants under the most unfavorable circumstances."

James himself described several of the plants he collected, including blue columbine, *Aquilegia caerulea*, now the state flower of Colorado. Most of the specimens found their way to Torrey and other botanists, and many came to bear James's name—for example, James's saxifrage, *Telesonix jamesii*, snow-

Jamesia

lover, *Chionophila jamesii*, and at least a dozen others. James, with two companions, climbed Pike's Peak; they were the first to climb any of Colorado's fourteen-thousand-foot peaks. He had only a few uncomfortable hours at the summit, but he managed to make the first collection of alpine plants from the Rockies. Stephen Long named the mountain "James Peak," but that name gave way to that of its discoverer, Zebulon Pike. Another peak was later named for James.

When James joined the Long Expedition he was only twenty-three years of age and had had a rather cursory training in botany and medicine. At Middlebury College in Vermont he had taken courses in botany, and while studying medicine with his brother in Albany, New York, he attended some of Amos Eaton's lectures. It was Eaton who introduced him to John Torrey, and Torrey who recommended him to Stephen Long. The trip was the grand adventure of his life, and upon his return he was asked to edit the journals of the expedition, which were published in 1823. Although James lived for many years, he was never again involved in exploration or to any extent in botany. For a time he was an army surgeon at posts in Iowa and Michigan, but he later retired to a farm in Iowa and became something of a mystic. He died in 1861 after falling from a wagon and being run over by the wheels. It was an unfortunate end for a person whose career had begun so brilliantly. But he is well remembered by those of us who look forward each year to the blooming of *Jamesia* and a host of other plants that bear his name.

John Torrey (1796–1873) had received a medical degree and was a practicing physician in New York City, but his interests were primarily in botany. His father was a prison official, and John's interests had been kindled by one of the inmates, Amos Eaton, who shortly after his release also stimulated Edwin James's interest in botany.

Torrey accepted a professorship at Columbia in 1827, and three years later a professorship at Princeton; he held the two positions concurrently for many years. In 1832 Asa Gray (1810–1888), who had read Eaton's *Manual of Botany* and was teaching high school in upstate New York, came to the city to become Torrey's assistant, and the two collaborated on a *Flora of North America*. Later in Torrey's life, a group of his younger associates founded the Torrey Botanical Club, the journal of which was the first botanical monthly in the country. The club also had much to do with the founding of the New York Botanical Garden, where Torrey's library and herbarium were deposited.

Torrey and Gray were innovators in that they abandoned Linnaeus's system of classifying plants strictly by their sexual organs and used a broader spectrum of features in the effort to develop a less artificial system of classification. Asa Gray, who went on to a long career at Harvard, became a friend of Charles Darwin and the major advocate of Darwinism in America, while his formidable colleague Louis Agassiz firmly maintained that species were incapable of change once they had left the hand of the Creator.

Darwin had written Gray in 1857, two years before the publication of *The Origin of Species*. He feared that Gray, a faithful member of the Congregational Church, would consider his notions atheistic. But, wrote Gray to Darwin with a twinkle in his eye, "Almost thou persuadest me to have been a hairy quadruped of arboreal habits, furnished with a tail and pointed ears." When the *Origin* appeared, Gray wrote a long, cautious, but favorable review in the *American Journal of Science*. This was followed by a series of more popular articles in the *Atlantic Monthly*. Agassiz remained unconvinced and until death held that the theory of evolution was "mischievous." "I trust to outlive this mania," he wrote. But it was Gray who ultimately prevailed.

For many years Gray and his wife lived in a house in the Harvard Botanical Gardens once occupied by Thomas Nuttall.

(Gray had met Nuttall before the latter moved to England, but had developed a dislike of him, perhaps because of his eccentricities.) Gray was one of the founding members of the National Academy of Sciences and a member of over sixty scientific societies throughout the world. He was a lucid and prolific writer. His *Manual of Botany* became the standard book for identifying the plants of the northeastern United States. It has gone through numerous editions and remains one of the most-used books in my library.

Both Torrey and Gray were mild-mannered, studious, and friendly persons, with good senses of humor—somehow attributes of all good botanists. Both did much to encourage other, younger botanists. They have been called "closet botanists," since most of their time was spent in herbaria and classrooms. But they did travel. In 1872 Gray went to California, where he spent a week with John Muir, who found him well able to keep up with him in the mountains. Afterward Muir sent him plants, one of which Gray named *Ivesia muirii*. A few weeks later Torrey visited Muir, and the two together effectively diverted Muir's attention from geology to botany.

During the same year the two botanists visited Charles Parry at his cabin in Colorado. Parry had recently named twin peaks Gray's Peak and Torrey's Peak. Gray and his wife, along with Parry and several others, climbed Gray's Peak and held a celebration on the top. Coming along a month later, Torrey stood at the foot of the peak named for him, but he was seventy-six years of age and too feeble to climb it. His daughter, who was with him, did however climb Gray's Peak. A year later Torrey was dead, a sprig of *Torreya* (a yew) on his casket. Torrey pine (*Pinus torreyana*), a dramatically shaped tree with huge cones weighing a pound or more, is a more commonly known reminder of this great botanist. Named by Parry, it grows in the wild in a restricted area along the windswept California coast, and is sometimes grown as an ornamental.

In 1877 Gray and his wife, accompanied by English botanist Sir Joseph Hooker, botanized across the Rockies and Sierra. At Mount Shasta, Gray once again met John Muir and asked him why *Linnaea borealis* (twinflower) had not been found in northern California. According to Muir, "Gray had felt its presence . . . on the mountain ten miles away." The next day they found it.

Gray and his wife made several trips to the Southeast in the late 1870s, finding *Torreya* in its native habitat and searching for *Shortia*. Early in his career, in Paris, Gray had found a plant collected by André Michaux in the mountains of the Southeast many years earlier and never identified. Gray considered it to belong to a new genus, which he called *Shortia*, after Kentucky botanist Charles Wilkins Short. For many years the plant could not be found in the wild, and botanists began to question Gray's identification of the plant as part of the American flora. Finally it was rediscovered by an herb collector in a remote part of the Carolinas, but the Grays arrived too late to find it in bloom.

On his seventy-fifth birthday, Gray received letters of congratulations from 180 American botanists as well as a silver vase embossed with figures of some of the plants he had studied. In 1887 he and his wife made a final visit to Europe, where he received honorary doctorates from Cambridge, Oxford, and Edinburgh. There were visits to Kew Gardens and to Mrs. Darwin. Back home, one of Gray's final acts before the stroke that eventually proved fatal was his signing of copies of his review of Darwin's *Life and Letters*.

Like Torrey, Gray had his genus, *Grayia*, a hopsage. It is a homely enough tribute to a man who not only contributed greatly to the classification of plants but also helped to nurture the beginnings of evolutionary biology in America and, by pointing out the similarities between the plants of eastern North America and those of East Asia, helped to found the science of biogeography. Perhaps the greatest tribute to these two giants of American botany are Gray's and Torrey's peaks, both over fourteen thousand feet in elevation and in summer sparkling with alpine flowers.

Jeffersonia *(Twin-leaf)*

When Edwin James completed the editing of the journals of the Long Expedition in 1823, Thomas Jefferson (1743–1826) had been retired and living at Monticello for more than a decade, after a long career as a leading spirit in the expansion of America's scientific vision. No doubt he read James's report eagerly, as he himself had been a major exponent of westward expansion. He had assisted André Michaux in planning his trek west in 1793, urging him to go as far as the Pacific (Michaux only went as far as the Mississippi, but that was an accomplishment for the time). Even before the purchase of Louisiana had been finalized, Jefferson was planning an expedition that in 1804 materialized as Lewis's and Clark's Corps of Discovery.

Early in his career, while recuperating from a fall from a horse, Jefferson had written *Notes on the State of Virginia*, a wide-ranging review of the geography, resources, and natural history of his native state. He listed 130 native plants and 77 birds but added little to what was already known. The most memorable section of his *Notes* was that in which he took issue with the great French naturalist Comte de Buffon. In his monumental *Histoire Naturelle*, Buffon had painted a dismal picture of North American wildlife. "The animals of America are tractable and timid [he wrote]. All animals are smaller in North America than in Europe. Everything shrinks under a niggardly sky and unprolific land." Jefferson countered with some figures: one American bear weighed 410 pounds, compared to the largest European bear, which weighed in at 154 pounds; the American beaver outweighed the European beaver 2:1; and so forth. To top it off, Jefferson had a New Hampshire moose sent to Buffon, who was impressed but never corrected himself in print.

Jefferson arrived at his inauguration as vice president with the bones of a giant animal which he called *Megalonyx* (Greek for "giant claw"). He believed it to be a huge carnivore, but it turned out to be a giant ground sloth. Later, in the presidential mansion, there was a pet mockingbird and a steady stream of visiting naturalists. On one occasion the floor of the White House was strewn with over three hundred fossil bones that had been quarried from a bone lick at the president's own expense. Jefferson's actions in the political arena were often criticized, and when he left office he confessed to a friend that "the whole of my life has been at war with my natural tastes, feelings and wishes. . . . [L]ike a bow long bent I resume with delight the character and pursuits for which nature designed me."

In his later years, Jefferson devoted much of his attention to the University of Virginia, which he had founded and hoped would be a means of transmitting his own understanding of nature to future generations. With so many memorials to Jefferson—the Declaration of Independence, the Louisiana Purchase and exploration of the West, the University of Virginia, and of course the imposing Jefferson Memorial in Washington—it is perhaps inconsequential to mention that in 1792 Jefferson's friend Benjamin Smith Barton named a plant *Jeffersonia*—an early spring flower of eastern forests bearing white flowers and large leaves that are divided in two, rather like the wings of a butterfly (hence the common name twin-leaf). Yet it is good that Jefferson's name lives in nature, aloof from the human drama that so often diverted his attention from things of the earth that he so loved.

Kirtland's Water Snake

$\mathcal{W}$ater snakes are placed in the genus *Natrix*, a Latin word related to the Greek word *nektris*, a swimmer. Kirtland's water snake (*Natrix kirtlandi*) is the least aquatic of the water snakes, preferring damp fields, marshes, and streamsides, where it can sometimes be found by turning over stones and logs within its range, the Ohio Valley and lower Great Lakes. Smallest of the water snakes, it is no more than fifteen to twenty inches long when fully grown. It is a perfectly harmless snake, but when disturbed it will flatten its body, ribbonlike, and strike feebly. In late summer the female gives birth to half a dozen or more living snakelets. Kirtland's water snake has a reputation of being a "city snake," since it is often found in city parks or suburbs. Perhaps that is a good place to locate its favorite food, earthworms, but it is a precarious habitat, living so close to bulldozers.

Needless to say, snakes are not the most beloved of animals, but they do have their advocates, none more devoted than Albert and Anna Wright, whose *Handbook of Snakes* is filled with intimate detail on even the least of these creatures. Their description of the colors of Kirtland's water snake is enough to send one out looking for one immediately. The spots on the back, they say, are "black, brownish olive, warm sepia, or burnt umber." The throat is "almost pure light buff, pale pinkish buff, or cream-buff," passing into "apricot orange, rufous, flesh, ocher, vinaceous-cinnamon, or orange cinnamon"; still further back it grades into alizarine pink. If this description seems overdone, it is only because the Wrights were meticulously following Robert Ridgway's *Color Standards and Nomenclature.* I knew the Wrights well. The basement of their home in Ithaca, New York, was filled not with jellies, jams, pickles, and other conventional preserves, but with jars of alcohol and formalin containing snakes and frogs collected by themselves and gleaned from others. Theirs was a labor of love, supported mainly by their own limited income. By their time (they were most active in the 1930s through the 1950s) all the North American snakes and frogs had been described, so they left their names attached to none of their favorite animals. But they were aware that naming is only the first step to knowledge, and they saw the need to make the findings of naturalists available to all through a series of handbooks on diverse groups of often neglected "lower" animals.

"Human beings have an innate fear of snakes," wrote E. O. Wilson in his book *Biophilia,* "or, more precisely, they have an innate propensity to learn such fear quickly and easily past the

Kirtland's Water Snake

age of five." Repellency to snakes, he says elsewhere, is "built into the brain in the form of a learning bias [to] become alert quickly to any object with the serpentine gestalt." In other words, snakes by their nature produce a biophobia that runs counter to our ability to affiliate with nature (biophilia). I do not believe this. Spencer Baird encouraged his daughter Lucy to play with snakes (harmless ones, of course), and he demonstrated to his own satisfaction that children have no trouble relating to snakes. My own grandchildren delight in catching garter snakes and even give them names before releasing them to go about their business. In temperate climates, it is easy to learn to recognize the few poisonous snakes, or other dangerous animals, and to treat them with due caution. We do not, after all, fear our automobiles, which are far more deadly than anything in nature.

Kirtland's water snake is, in any case, so small and secretive that it is known only to a few. Jared Potter Kirtland (1793–1877) was a native of Connecticut, trained in medicine at Yale and imbued with a love of nature through his grandfather, also a physician. When he was thirty he moved to Ohio, where he practiced medicine and later became one of the founders of the medical branch of what is today Case Western Reserve University. On the side he experimented in floriculture and became an expert taxidermist. He accumulated a large collection of birds that was later incorporated into the Cleveland Academy of Natural Sciences (later the Cleveland Museum of Natural History), which he helped to found. The scientific publication of the museum was later named for him: *Kirtlandia.* One of Kirtland's biographers wrote of his "universal and unextinguishable pursuit of knowledge and an enjoyment of nature which kept him fresh and green and youthful to the very last."

Kirtland had a special interest in mollusks, and it is distressing to learn that, as early as 1851, population growth and industrial development were raising havoc with mollusk populations. On January 11 of that year he wrote to Spencer Baird:

Localities that 10 years since abounded with the finest specimens of [bivalves] are now entirely barren, and some species
have become nearly or quite extinct in the Cuyahoga River.
Both fluviatile and terrestrial univalves have also decreased
with equal rapidity. The delicate *Helix Sayii* is no longer
found in Northern Ohio so far as my observation extends,
and many other species that were once very common are now
to be met with only accidentally.

Besides the snake, Kirtland's warbler (*Dendroica kirtlandii*) was
named for him, in this case by Baird on the basis of specimens
Kirtland had collected near his home. This is one of the rarest
of warblers and is now listed as endangered. In its summer
range it is confined to north-central Michigan, where it breeds
in thickets of jack pine. Controlled planting and burning are
now being conducted in the hope of maintaining fragments of
a habitat that was once more widespread.

Robert Kennicott (1835–1866) was the discoverer of Kirtland's water snake in 1856, not far from his home near Chicago.
His dedication was as follows:

In giving to this serpent the name of Dr. Kirtland, as a slight
token of the respect due him, to whose enthusiastic and
untiring devotion to Science the West owes so much, I would
also make some expression of my personal gratitude to the
honored teacher, whose kind encouragement and instruction
led me to study Nature, by dedicating to him his pupil's first
contribution to Science.

Kennicott was brought up in northern Illinois and spent much
of his life there. As a youth, his health was too frail to enable
him to acquire a formal education, and he spent much time
wandering the local woods and fields. He studied for a time
with Kirtland and corresponded with Baird, whom he met in

Washington in 1854. Baird welcomed him into his home. He "was destined to add, by his travels and collections, directly and indirectly perhaps more than any other collaborator, to the riches of the Smithsonian collection," wrote W. H. Dall in his biography of Baird. Back in Chicago, when only twenty-one, he helped to found the Chicago Academy of Science, which became the focus of much of his life.

At Baird's instigation, Kennicott participated in an expedition of three years' duration into northern Canada. Here he lived with members of the Hudson's Bay Company, whom he taught to collect birds and make skins. A stream of specimens flowed to Chicago and Washington, many of them representing new or little-known species. After spending time in Chicago and Washington studying his material, he was once again lured to the Far North, this time on an expedition launched by the Western Union Telegraph Company to survey a possible route for a telegraph line. W. H. Dall was also a member of this expedition. But one morning a member of the party found Kennicott dead of a heart attack on an Alaska beach. A letter to Lucy Baird told of the loss of "the jolly and warmhearted and zealous Kennicott." He was only thirty-one.

Kennicott's discoveries included subspecies of the arctic ground squirrel, the arctic warbler, and the screech owl, all named for him. He was also an avid collector of insects: I know of at least five species that were named for him. One of them, *Cerceris kennicotti*, is a ground-nesting wasp that is common around my home, serving as a reminder of the short life of a keen and exuberant naturalist.

LeConte's Thrasher

*L*eConte's thrasher (*Toxostoma lecontei*) is a bird of the hottest of deserts, that of the lower Colorado River basin in Arizona and California and of the Mexican states of Sonora and Baja California. This is the desert that John Alcock has described so eloquently in his books *Sonoran Desert Spring* and *Sonoran Desert Summer.* Daytime temperatures in the summer often exceed one hundred degrees for weeks on end, and the mean annual rainfall is barely five inches. As Alcock says, only very special plants and animals can thrive in such a hostile environment. Those elegant, long-tailed vocalists, the thrashers, do well here; in fact, half of North America's eight thrasher species are largely confined to the Sonoran desert. All four have long, down-curved bills, which explains the generic name *Toxostoma,* based on the Greek words *toxon,* bow, and *stoma,* mouth. These scimitarlike bills help the birds thrash about in ground litter in the search for insects (which I suppose is the origin of the word *thrasher*). These birds (along with the mockingbird and the catbird) make up the family of mimic thrushes (Mimidae), all of them with rich and varied songs that often contain phrases copied from other birds.

LeConte's Thrasher

LeConte's, sometimes called the desert thrasher, is the rarest of its clan and also the most fully adapted to hot and sandy desert soils. The bird is itself more or less sand-colored, except for a long, black tail that is cocked high as it runs about the desert floor. When flushed, these birds prefer to run, keeping the thickest cover between them and the intruder; but if pursued, they will fly low over the desert, weaving a tortuous course among the bushes. At morning and dusk, in the mating season, the male sings an exuberant song from a bushtop, a song that can sometimes be heard nearly a mile away. Hottest parts of the day are doubtless spent resting in the shade, though in fact no one has worked out the time budget of these elusive birds. The usual habitat is flat, sandy desert or sandy washes, where there are creosote bushes and cholla cacti. Nests are built deep within the spiny branches of the cacti, where only the toughest and most persistent of predators is likely to find the eggs or nestlings. There are three or four eggs, greenish-blue spotted with brown and lavender. The nestlings have short, straight bills, but as they grow the bill becomes gradually longer and more down-curved.

It was John Lawrence LeConte (1825–1883) who discovered these birds near Yuma, Arizona, in 1843, when he was a young man exploring the West on his own initiative. He had traversed the Santa Fe Trail and gone on to California, having a number of adventures on the way. While he was traveling along the Gila River, Indians stole his horses, and he had to hike thirty miles across the desert to the nearest settlement. This was the first of many trips LeConte was to make throughout the country as well as in Europe and North Africa, primarily in search of beetles, his major passion.

LeConte was a member of a wealthy and prominent Huguenot family that had fled to America around 1700. His father, Major John Eatton LeConte, served for a time in the army but devoted most of his life to the study of plants, insects, and reptiles, and he taught his son to collect and draw specimens for him. Father and son were constantly together, the major's wife having died a few months after her son was born. According to Robert Dow, the major worked at his desk on his specimens with "the little toddler on his knee." John LeConte the physicist and Joseph LeConte the geologist were cousins of John Lawrence LeConte, whose middle name is usually used or abbreviated to avoid confusion with his father and his cousin.

J. L. LeConte attended Mount St. Mary's College, and the story is told that one day, in class, he leaped from his seat and scrambled about the floor. He was reprimanded and had to explain that he had seen and captured an unusual beetle that was crawling about the classroom. Later he attended medical school, though he never established a practice. Father and son settled in Philadelphia, where both became active in the recently founded American Entomological Society. During the Civil War, J. L. LeConte was a medical officer with the rank of Lieutenant Colonel, and after the war he served for a time on a railroad survey in Kansas and New Mexico. In the meantime he had been working on a comprehensive classification of the beetles of North America, which he completed with the help of his younger colleague George H. Horn. Earlier, LeConte had rendered a major service by gathering the scattered publications of pioneer entomologist Thomas Say and publishing them in two volumes, including the color plates from Say's *American Entomology*.

LeConte's publications spanned forty years; some were in the fields of geology, fossil mammals, and ethnology, but by far the majority were on beetles. He purchased specimens from John Abbot, John Xántus, and many others. His huge beetle collection was left to the Museum of Comparative Zoology at Harvard, presumably because of his long friendship with Louis and Alexander Agassiz. LeConte was one of the founders of the National Academy of Sciences and a president of the American Association for the Advancement of Science. Despite his wealth and prestige, he was approachable and generous. In an obituary, George Horn wrote: "We all knew him as a cultured scholar, a refined gentleman, a genial companion, a true friend."

LeConte described over five thousand species of beetles and had many named in his honor. It is perhaps ironic that the only vernacular name attached to any of the species he discovered pertains not to a beetle but to a bird. The LeContes were distantly related to Spencer Baird, and all of them, J. L., John, and Joseph, collected birds for the Smithsonian. (John, the physicist, also had a bird named for him: LeConte's sparrow.) Writing to his brother William concerning the receipt of bird skins from J. L. LeConte, Spencer Baird added: "Don't forget about getting bugs for him." It is remarkable how well most naturalists of the nineteenth century knew one another and were helpful to one another.

LeConte's thrasher was described by George Newbold Lawrence (1806–1895), a friend of Audubon and one of the leading ornithologists of his time. He was a native of New York and like so many naturalists became interested in birds as a youth. Under the influence of Audubon and of Baird, he left the family business in midcareer and devoted the rest of his life to the study of birds. He did not do very much field work himself but was sufficiently well-to-do to be able to assist the Smithsonian Institution in launching several expeditions for collecting specimens. Lawrence specialized in birds of the American tropics. His collection of over eight thousand bird skins was deposited in the American Museum of Natural History in New York. He described over three hundred species of birds and had several species and a genus named for him. Nearly all were tropical, but a California finch, *Carduelis lawrencei*, Lawrence's goldfinch, bears his name. Lawrence's warbler, *Vermivora lawrencei*, combines the features of the blue-winged and the golden-winged warblers and is now regarded as a rare hybrid between those two species. Many years ago I was lucky enough to see one in a Connecticut woodland. At the time, Lawrence's name was unknown to me, but like any budding naturalist I was excited to see anything the books described as rare.

Lincicum's Ants

Gideon Lincicum (1793–1874) had, like J. L. LeConte, an interest in medicine and in insects, but aside from that it would be hard to find two individuals more unlike. LeConte came from a wealthy and distinguished family and had the best education available. Lincicum is best described as a curious backwoodsman—curious about things around him and a curiosity himself. He was born in Georgia, son of an itinerant preacher, and spent much of his early life in Mississippi. He had only a few months of formal schooling, but enough in those days to qualify him as a teacher; later he "read medicine" enough to hang out a shingle as a frontier physician. In 1835 he visited Texas, still part of Mexico, where he was captured by Comanches. He claimed to be a medicine man and persuaded the Comanches to let him leave camp to bring back a powerful medicinal plant. "I rode slowly away until I got out of sight [he wrote in an *Autobiography*], and then, changing my course, rode rapidly all that night, all the next day, and until 12 o'clock the day following."

In spite of this experience, in 1848 Lincicum moved to Texas permanently and decided to devote the rest of his life to natural history and to enjoying his violin. He was irreverent and sometimes uncouth, but he had an intense curiosity about nature and was bold enough to write to Louis Agassiz and to Philadelphia naturalists E. D. Cope, E. T. Cresson, and Elias Durand. He began to collect plants for Durand and insects for Cresson. Before he went to Mexico in 1867, he sent a collection of birds, mammals, insects, fossils, and shells to the Smithsonian Institution. A friend later found "old Gid" deep in Mexico, barefooted, playing his fiddle and "communing with nature." He had requested that his fiddle be buried with him.

Texas is alive with ants, and Lincicum often spent hours watching them. He was convinced that harvester ants plant certain kinds of grasses around their nests, harvest the grains for food, and even weed their gardens. He sent some of his observations to Charles Darwin, who communicated them to the Linnaean Society of London. His remarks were met with skepticism (of course the grasses growing around the nests result from the fact that the ants discard stored seeds that start to sprout). But Lincicum went further, discussing the organization of ant colonies as if the ants were rational beings. Some ants, he said, were intellectually superior to others and became the leaders. A Philadelphia minister, H. C. McCook, became interested in Lincicum's observations and left for Texas to check them out. In his book *The Natural History of the Agricultural Ant of Texas*, McCook corrected some of Lincicum's more extravagant interpretations.

Lincicum also attracted the attention of Texas's state geologist, S. B. Buckley, who published descriptions of seventy-six "new species" of ants. Harvard's professor William Morton Wheeler later studied Buckley's descriptions and found them "fearfully and wonderfully made"; sometimes Buckley even misidentified the sexes. Most of Buckley's names have been relegated to the scrap basket, since his "species" cannot be identified; this includes three that he named *lincicumi*. The name of the post oak grape, *Vitis lincicumii*, also described by Buckley, is still in use, however.

Lindheimer's Daisy

It seems odd that Lincicum sent his plants to Durand rather than to the more eminent botanists Engelmann and Gray, who would have made better use of them. It is also odd that he had little or no contact with another plant collector active in Texas at the same time. This was Ferdinand Jakob Lindheimer (1801–1879). Lindheimer was born in Germany and doubtless spoke German most of his life, so he and Lincicum may have had difficulty communicating if they were acquainted. Lindheimer had moved to America in 1827 and settled in a German community in Illinois before moving to New Orleans and then to Vera Cruz, Mexico. He began to collect plants and send them to his good friend George Engelmann, and after moving to Texas at the time of the defeat of Santa Anna's forces, he spent a few months working with Engelmann in St. Louis. Engelmann wrote to Asa Gray asking if he would like someone to collect plants for him in Texas. Gray responded with an offer to pay Lindheimer eight dollars per hundred specimens. Lindheimer accepted and spent several years roaming the eastern half of the state and shipping packets to Gray. A friend described how he went about his collecting:

> He bought a two-wheeled covered cart with a horse, loaded it with a pack of pressing-paper and a supply of the most indispensable provisions, namely flour, coffee, and salt, and then set forth into the wilderness, armed with his rifle and with no other companion than his two hunting dogs, while he occupied himself with collecting and pressing plants. He depended for his subsistence mainly upon his hunting, often passing whole months at a time without seeing a human being.

In 1852 Lindheimer settled in New Braunfels, where he edited the local German newspaper for nearly twenty years. He no longer collected plants, but he had already collected so many that his name has become affixed to at least twenty Texas plants by my count. Asa Gray and George Engelmann honored him by naming a daisy *Lindheimera texana*. Called Lindheimer's daisy or Texas yellow star, the brilliant flowers of these plants are spangled in summer across East Texas fields, often intermingled with bluebonnets. The grass *Muhlenbergia lindheimeri* combines Lindheimer's name with that of an eminent grass specialist who will appear in the next chapter.

Merriam's Mouse

*R*odents are the most successful and widespread of all mammals, and even the most fastidious housewife is likely to keep a trap or two on hand in case of invasion by house mice (which like Norway rats are not natives but came over on the Mayflower, or probably earlier). Few people are aware that there are a great many native species of small rodents, most of them strictly nocturnal and therefore rarely seen. Indeed, even early naturalists were largely unaware of this until the invention of the spring trap, about 1887. It was "the trap that made Merriam famous," for C. Hart Merriam (1855–1942), then head of the Division of Economic Ornithology and Mammalogy of the U.S. Department of Agriculture (later the Biological Survey), devoted much of his life to the sudden deluge of previously unknown mammals. He and his colleagues set out baited traps all over the continent and began to describe new species and subspecies furiously. Merriam's mouse (*Peromyscus merriami*) is only one of many.

Members of the genus *Peromyscus* are not the mundane, gray and dust-covered mice well known to homeowners; they are country mice, not city mice. They are attractive sprites, with big ears, shining eyes, and long, slightly hairy tails. Their feet and bellies are white, hence the common name white-footed mice. Deer mice is a more commonly used name, though less easily explained. More than a dozen species range throughout North America. When pioneers built their cabins in the wilds, deer mice often made their appearance and, if well treated, sometimes became rather tame, helping to pass long winter evenings. There is hardly a woodland, field, or even a desert, where these mice do not live and carry out their nightly excursions for seeds, buds, and other plant materials. Several kinds of owls would be hard put to survive if they could not depend upon a good supply of deer mice.

Merriam's mouse is a desert species, ranging from north-central Mexico into southern Arizona, chiefly where there are dense stands of mesquite (mesquite mouse is an alternate name). It has never been well known, and with mesquite being cut for firewood or destroyed to make more (very marginal) land available for cattle grazing, its future is uncertain. A number of other small rodents occur in this same habitat: cactus mice (also a *Peromyscus*), desert pocket mice, Merriam's kangaroo rats, white-throated wood rats, and yuma antelope squirrels. It is surprising that so many species can make a go of it where food resources can surely not be plentiful.

Merriam's Mouse

Because of its limited distribution, Merriam's mouse has attracted little attention from biologists. But deer mice, as a group, have proved extremely useful for demonstrating various life phenomena. They are easy to catch in spring traps or live traps, and most species do well in captivity. So many studies have been made on these small rodents that two recent books have been produced that review some of this research. Either one is weighty enough that if dropped on a deer mouse it would mean certain death. (But given access and time, the mice could readily chew them into enough pulp for many nests.) There is also a *Peromyscus Newsletter*, published at the University of South Carolina, which also maintains stocks of several species and several mutant stocks, available to researchers. Clearly these mammals have excited biologists all out of proportion to their size.

One reason for their popularity is their ability to produce several generations in a year, in the field as well as in the laboratory. Females become ready to mate when four to seven weeks old (based on studies of several species, but not *merriami*, which as I have said has been little studied). Usual litter size is two to five; the young are born blind, hairless, and toothless and weigh only one or two grams. They are weaned when three or four weeks old, and soon thereafter they leave the nest. Deer mice sometimes live several years in laboratory cages, but it is doubtful that many live for more than a year in nature.

People who have followed natural populations by live-trapping year after year have usually found deer mice populations to be relatively constant, without the wild fluctuations of some other rodent populations (lemmings, for example). So there appears to be some natural mechanism for preventing overpopulation. The mice tend to have rather ill-defined home ranges which they maintain by mutual avoidance and occasional scrapping, but the ranges are smaller when populations are higher, so this seems an inefficient population control. Very probably crowding influences the production of sex hormones. It is hard to study this in the field, but females in crowded laboratory colonies have been shown to have a much lower pregnancy rate than those reared alone with a male.

When young deer mice disperse from the nest where they were born, the males travel much farther than do the females. This reduces the possibility of a male mating with his sisters or his mother. When brothers and sisters from the same litter are allowed to mate after reaching maturity, their litters are smaller and the young lighter in weight (as compared to the offspring of unrelated individuals); there is also more infant mortality. This is called inbreeding depression and is similar to that occurring in many animals, including humans. Differential dispersal of the sexes as they leave the nest is part of the animals' means of avoiding inbreeding. The dire results of incest are well known in humans. E. O. Wilson, in his monumental book *Sociobiology,* cites an example from Czechoslovakia. Of a sample of 161 children born to women who had sexual relations with their fathers, brothers, or sons, 15 died at or shortly after birth and nearly half had serious physical or mental defects. In a similar group of children born to the same women through nonincestuous relationships, few died as infants or had abnormalities.

Deer mice are also able to assess degree of relationship, through odor stimuli, just as we are able to do through knowledge of kinship. Brian Keane of Purdue University supplied female white-footed mice that were ready to mate with a choice of nest bedding: some unsoiled, some soiled by a brother, some by an unrelated male, and some by a first cousin. Females showed a preference for the odor of a first cousin. They also showed more friendly behavior and less aggression toward first cousins than toward brothers or unrelated males. While avoiding close inbreeding, females do prefer males with some relatedness. This may reflect an avoidance of "outbreeding depression," that is, the risk of losing adaptation to local conditions by mating with an individual of low genetic similarity. Again, an analogy is possible with human behavior, as we, too, generally select mates from our own social milieu.

Much of the research on deer mice (only a tiny fraction of which is reviewed here) deals with adaptive differences among the several species. Merriam's mouse may hold some surprises when it is finally studied in detail, since it is strictly a desert species. Merriam himself was, of course, unconcerned with the life-styles of the small mammals recovered from his traps. He lived at a time when the discovery and naming of "new species" was all-important. Indeed he was notorious for describing species on the basis of minute differences in the color of the fur, length of appendages, and the like; he was a "splitter."

Merriam and Theodore Roosevelt were good friends, but they did not always see eye to eye concerning animal classification. Roosevelt attacked Merriam in the pages of the journal *Science* for having split the coyote into seven species and the grizzly bear into eight. "The excessive multiplication of species based upon trivial points of difference [wrote Roosevelt] merely serves to obscure the groupings which are based on differences of real weight." But when Merriam named the Roosevelt elk (*Cervus roosevelti*) from the Northwest, Roosevelt was "more pleased than I can say. . . . To have the noblest game animal of America named after me by the foremost of living mammalogists is something that really makes me prouder than I can say." But the Roosevelt elk, too, was based on trivial differences from other elks and is now rated at best a weak subspecies.

Merriam is perhaps best known as the originator of Merriam's life zones, a result of his study of small mammals from all over the continent. These zones were established on the basis of temperature and rainfall and appear to limit the distributions of many animals. They are no longer accepted as widely as they once were, but some of the terms Merriam invented are still used frequently: for example, "transition zone" for a strip across the continent transitional between a more boreal climate and a warm temperate climate; and "lower Sonoran zone" for areas relatively frost-free and with an average annual rainfall of under ten inches.

As head of the Biological Survey, Merriam was much concerned with the enforcement of game laws, such as they were in his time. He was incensed by a Pennsylvania law that offered a bounty of fifty cents on hawks, owls, weasels, and minks. In a year and a half this resulted in the killing of 128,571 predators at a cost to the state of $90,000. The total saving to farmers, he calculated, amounted to $1,875. "But," he added, "every hawk or owl destroys at least a thousand mice or their equivalent in insects. . . . The slaughter of such vast numbers of predaceous birds and mammals is almost certain to be followed by a correspondingly enormous increase in the numbers of mice and insects formerly held in check by them. . . ."

The importance of predators in natural ecosystems is a lesson that does not seem to have been learned in the century following Merriam's remarks. The U.S. Department of Agriculture's Animal Damage Control Program in 1988 saw to the destruction of 76,033 coyotes, 1,163 bobcats, 4,427 red foxes, and 203 mountain lions, thus saving the lives of several lambs and calves at an annual cost of $30 million. That coyotes, bobcats, foxes, and mountain lions are infinitely more admirable than cows or sheep is, of course, not a matter worthy of attention in our dollar-driven society.

Merriam was at times a difficult person, and he loved to tell off-color stories. The story is told on him that he was once stung by a velvet ant (a wingless wasp, sometimes called "cow killer"). Thereafter, whenever he saw one, he would dismount from his horse and urinate on it in revenge.

In 1884 he hired a relatively uneducated assistant, Elizabeth Gosnell, to skin animals, type his notes, and do other odds and ends. Over time "Lizzie" improved her education and became indispensible to him, even accompanying him on harrowing field trips. They were married in 1886 and named their second daughter Zenaida, after a genus of doves that had in turn been named for the wife of ornithologist Charles Bonaparte.

In the course of time Merriam had a shrew, a chipmunk, a pocket mouse, a kangaroo rat, a lizard, a pupfish, and even a louse named for him. It was Edgar Alexander Mearns (1856–1916) who named the deer mouse. Mearns and Merriam had shared living quarters when they were medical students in New York, but Mearns chose to follow a career as an army surgeon rather than becoming a professional biologist. He had been an ardent naturalist and collector from youth, and while in the army he sought assignment to places of much biological interest. In 1891 he was appointed medical officer to the Mexican-American International Boundary Commission. He collected some thirty thousand specimens for the U.S. National Museum and published a monograph on the mammals of the Southwest. In contrast to Merriam's somewhat prickly disposition, Mearns was a modest and unassuming individual, rarely critical of others and generous with his specimens. With Merriam, he was one of the founders of the American Ornithologists' Union in 1883.

Beginning in 1903 Mearns spent several years in the Philippines, where he collected many plants, birds, and mammals for American museums. In Manila he found that a bounty had been placed on rats, since they were vectors of plague. He asked for a sample and was presented with several garbage cans full of dead rats. It was assumed all were common "ship rats," but in fact Mearns extracted several new species from the lot. Over time Mearns ascended several of the highest peaks in the Philippines, climbing through dense and uncharted forests during interminable rains. During an expedition against the wild Moro tribesmen, he often left the column of troops to hunt birds. "It was zeal and not ignorance of the risks [wrote a member of the party] that impelled Mearns to such incredible acts of heroic folly."

When Theodore Roosevelt made a hunting and scientific expedition to Africa in 1908, he took along Mearns, whom he characterized as "the best field naturalist and collector in the United States." Mearns and a companion climbed Mount Kenya, where they collected "1,112 birds, of 210 species . . . 1,320 mammals and 771 reptiles" (according to Roosevelt, in his book *African Game Trails*). In 1911 Mearns went on a second trip to Africa, this one with the expedition of Childs Frick. In spite of declining health, Mearns collected 5,200 birds and filled several notebooks with data. "Probably no more indefatigable collector ever roamed on African soil," reported Herbert Friedman, curator of birds at the U.S. National Museum.

Friedman compiled a list of birds described by Mearns; there are over 150, all from the Philippines and Africa. Over time, Mearns had many species and subspecies named for him, as well as three genera: *Mearnsia*, a tree of the myrtle family; *Mearnsia*, a swift (it is allowable to use the same generic name for a plant and an animal); and *Mearnsella*, a genus of fishes (all three were based on his collecting in the Philippines). He died, like every good naturalist, surrounded by notes and specimens so numerous that even another life would not have been time enough to do justice to them.

MacGillivray's Warbler

When Audubon named a warbler for William MacGillivray (1796–1852) he was expressing thanks for jobs well done. Audubon was most at home with a paintbrush; by his own admission he was a poor writer in English and not much better in his native language, French. On a visit to Scotland in 1830, he met MacGillivray, a professor at Edinburgh University who was an experienced naturalist and writer and willing to help Audubon write up his notes in what he called his *Ornithological Biography* (and later his *Synopsis*, to accompany his *Birds of America*). MacGillivray's fee was modest, two guineas per galley of sixteen pages. Despite other responsibilities, and despite criticisms of Audubon then rife, MacGillivray was faithful and diligent. "I have not seen so excellent a friend," he wrote Audubon. He was, however, sometimes annoyed that Audubon did not acknowledge his help to the extent that he felt he deserved.

MacGillivray had a distinguished career as a naturalist in his own right. His eight-volume *A History of British Birds* paid far more attention to the anatomy and life history of birds than was usual for its time. He was a busy lecturer and editor and published on topics in botany and geology as well as on birds. Exhausted from his labors, in 1850 he spent a month hiking in the Scottish Highlands, where he suffered from exposure to the extent that he never fully recovered, and he died two years later.

MacGillivray's warbler is little known except to westerners who are persistent enough to track a brilliant songster into the thickets where it lives. Surely so clear and bold a song—chiddle-chiddle-chiddle, turtle-turtle, in Roger Tory Peterson's interpretation—must belong to something rather special. Of course everything living is special, but these warblers are more voice than substance. Still, it is rewarding to an avid birder to spot the bird flashing its gray and yellow plumage in the midst of a tangle of willows and alders.

Menziesia *(False Azalea)*

Botanists Archibald Menzies and Gotthilf Heinrich Ernst Muhlenberg were born within a year of one another but had little in common geographically or in their backgrounds. Menzies (1754–1842) was a Scotsman who studied medicine and botany at Edinburgh University. As a young man, he traveled around the world as a surgeon on the *Prince of Wales*, engaged on a fur-trading expedition, and he took every opportunity to collect seeds and plants for his British correspondents. On his return the influential naturalist Sir Joseph Banks arranged for him to accompany Captain George Vancouver on a survey of the West Coast of North America. His duties were to collect seeds and plants for the Royal Botanical Gardens in Edinburgh and for Kew Gardens in London. He was also to distribute the seeds of fruit trees so that when British settlers arrived they would have fruit to harvest. In 1792, this was well before the West Coast explorations of Eschscholtz, Douglas, and Nuttall.

Menzies was on Vancouver's flagship, *Discovery*, and had frequent opportunities to go ashore, though usually not very far inland. His journals (later published in part) are filled with comments on the plants he saw and collected as well as descriptions of the scenery, which understandably greatly impressed the young naturalist. In the course of the expedition, several of the great volcanic peaks of the Cascades were named: Mount Baker, Mount Rainier, Mount St. Helens, and Mount Hood, all for British naval officers or other notables. Evidently it was a relatively uneventful expedition, with only minor problems from storms or illnesses and hardly any from Native Americans or the Spanish authorities in California. However, Menzies arrived in England under arrest for inso-lence and contempt as a result of a quarrel with Vancouver over the services of an assistant who had been assigned to Menzies to care for his plants. Menzies's influential friends intervened to secure his release.

Menzies was with Vancouver from 1791 to 1795. On his return he established a medical practice in London; aside from a short period in the West Indies, he remained there until his death at the age of eighty-eight. He published very little but was generous in providing data and specimens to others. Some of the plants he collected found their way to Frederick Pursh, others to his close friend William Jackson Hooker. David Douglas visited Menzies after Douglas's first trip to America and before his ill-fated second trip. Young Asa Gray visited him, finding him "a most pleasant and kind-hearted old man."

Menzies was the first to collect specimens of the coast red-woods, but it fell to another to name these magnificent trees after a Cherokee half-breed, Sequoia, who had invented the Cherokee alphabet. Several important West Coast trees and shrubs that Menzies collected do bear his name, including *Pseudotsuga menziesii* (Douglas-fir), *Arbutus menziesii* (coast madrone), and *Menziesia ferruginea* (false azalea). The last is a shrub two to six feet tall that provides a prominent understory in wet, coastal mountains. The leaves occur in whorls, suggesting azalea, but the flowers are quite different, being bell-shaped and of an unusual coppery color. Other names for the shrub are fool's huckleberry and rustyleaf. The leaves and fruits are poisonous to sheep.

Muhlenbergia *(Muhly Grasses)*

In contrast to Menzies, Gotthilf Heinrich Ernst Muhlenberg (1753–1815) was born in America and did little traveling. His father had come to Pennsylvania from Germany in 1742 as a leader in the Lutheran Church. Gotthilf (Henry to his friends) was trained in the ministry and became the pastor of the Lutheran Church in Lancaster, Pennsylvania. During the Revolution he had to flee into the country, and he used his idle hours to collect and classify plants. His father had advised him to "read the book of nature" but then became concerned that his son was more attentive to botany than to his parishioners. Muhlenberg identified more than a thousand species of plants around his home in Lancaster, and in 1791 sent his catalog of plants to the American Philosophical Society for publication. Muhlenberg then began a more general review of North American plants, but only the section on grasses was completed; it was published after his death. For much of his life he was a vigorous man and often walked the sixty miles between Lancaster and Philadelphia. He was universally admired, and his herbarium was meticulously maintained. His researches paved the way for John Torrey and Asa Gray, who were youths when he died.

Muhlenberg described over one hundred species of plants as well as the genus *Bartonia*, named for his Philadelphia friend Benjamin Smith Barton. Several plants bear his name, including chinquapin oak, *Quercus muehlenbergii*; a sedge, *Carex muhlenbergii*; and several mosses, lichens, and a fungus. *Muhlenbergia* is a large genus of grasses, nowadays usually called "Muhly grasses," a rather inelegant name by which to remember a Lutheran cleric who found time from his duties to render services to the science of botany. *Muhlenbergia* is, however, an important genus of grasses, with well over one hundred species, some of them growing several feet tall. Their blades and seeds provide food for a great variety of animals.

Nelson's Wood Rat

 ood rats are not the ugly, insidious rats that infest warehouses, dumps, and city sewers. These are outdoor rats, larger members of the same group as the deer mice and, like them, relatively big-eyed, big-eared, white-bellied, and long-tailed. Like most rodents they are creatures of the night, well known to the coyotes and owls that find them choice items in their diets. Collectively the species live in almost all natural habitats, from moist forests to mountains and deserts. All build fairly massive nests, in rock crevices, among boulders, around the bases of bushes, or occasionally in cabins or abandoned buildings. In the Southwest and in Mexico, they often build their nests in the ruins of ancient Indian villages.

Nests, or "houses," are built of almost anything they can collect within their ranges—sticks, cactus joints, bones, bits of dried animal dung, and various other odds and ends—hence the frequent name "pack rats." Houses usually contain a single wood rat, or a female and young, and are sometimes reoccupied generation after generation. In their book *The Dusky-footed Wood Rat*, Jean Linsdale and Lloyd Tevis describe and illustrate many of the houses of this species, some as much as two yards in diameter and two feet high. Usually there are several entrances. The nest proper is a bowl-shaped mass of fine, fibrous materials in the center. I have come to know wood rats rather too well, as each fall members of two species (Mexican and bushy-tailed) try to establish their houses beneath my house. They would be welcome if they were not so inclined to chew on electric wires. Fortunately they are easy to live-trap and move elsewhere.

Many people have commented on the diverse materials gathered by pack rats and incorporated into their houses or dumped just outside. In a nest of the white-throated wood rat in Colorado, Robert Finley found sticks of seven different kinds of trees and bushes, joints of several kinds of cacti, pieces of dung, bones, stones, yucca stalks, pine cones, owl pellets, feathers, and even part of the nest of a paper wasp. When wood rats nest in or near human habitation, they often pick up bits of glass, paper, or even silverware. Perhaps the most famous description of a pack rat's midden is that of Theodore Roosevelt, who had a ranch in the Badlands of South Dakota. "From the hole of one, underneath the wall of a hut, I saw

Nelson's Wood Rat

taken a small revolver, a hunting knife, two books, a fork, a small bag, and a tin cup."

Wood rats living in a cold climate gather food in the fall to carry them through the winter. The food consists of all kinds of plants, even formidable ones such as poison ivy, yuccas, and cacti, more often foliage than seeds. Mushrooms, grasses, and needles of conifers are also sometimes used. Robert Finley listed 219 kinds of plants used by the six species of wood rats that occur in Colorado. The rats do not hibernate, so depend upon their stores during cold weather. In contrast their distant cousins, the marmots and some of the ground squirrels, store fat in their bodies and sleep away the winter, a wholly different strategy that accomplishes much the same thing.

Clearly, foods stored by wood rats must be relatively nonperishable; otherwise the nest might be contaminated with molds and bacteria. When O. J. Reichman of Kansas State University gave caged eastern wood rats a choice of grapes or standard rat chow, they ate the grapes but stored much of the chow. Obviously they were able to discriminate between highly perishable food items and those that were more durable and to behave accordingly.

Edward William Nelson (1855–1934) collected the first specimens of Nelson's wood rat (Neotoma nelsoni) in the mountains of the state of Vera Cruz, Mexico, at an altitude of 7,800 feet. By Nelson's time, Merriam, Baird, Coues, and their collaborators had already done a good job clarifying the mammal fauna of the United States, but there were still parts of the continent that had been little explored. Nelson spent four years in Alaska and nearly fifteen years in Mexico before taking over Merriam's chair as head of the U.S. Biological Survey in 1916.

Since Nelson's wood rat is a subtropical species, it may not store food in its nest in the manner of temperate-zone species. It is fairly large, as wood rats go, but has been little studied since its discovery.

Nelson was born in New Hampshire but spent much of his youth in upstate New York and later Chicago. He loved the out-doors, and when a friend introduced him to Wilson's and Nuttall's books on birds, he began to collect birds and preserve the skins. While preparing a skin from a partially decayed speci-men, he contracted blood poisoning, and to help in his recovery he joined Edward D. Cope and others on a collecting trip to Wyoming, Utah, and Nevada. On his return he went to Washington to meet Spencer Baird, Robert Ridgway, and other naturalists. Through Baird's influence, Nelson joined an expedition to Alaska in 1877. In the summer of 1881 he was aboard the *Corwin,* which skirted the shore of Siberia in search of the lost ship *Jeannette,* which had disappeared two years earlier with DeLong's polar exploration party. His companion was John Muir, who wrote vividly of some of their adventures in his essay *The Cruise of the Corwin.* Muir and Nelson witnessed a slaugh-ter of walruses, which (in Muir's words) "lie on the ice without showing any alarm, waiting to be killed . . . oftentimes for their tusks alone." Scenes such as this stuck in Nelson's mind, and as head of the Biological Survey many years later he saw to the passage of the Alaska Game Law of 1925.

Nelson and Muir visited Saint Lawrence Island, in the Bering Sea, soon after many of the Eskimos had perished in the famine of 1878–1879. "The scene was indescribably ghastly and deso-late [wrote Muir], though laid in a country purified by frost as by fire. . . . The shrunken bodies, with rotting furs on them, or white, bleaching skeletons, picked bare by the crows, were lying mixed with kitchen-midden rubbish. . . . Mr. Nelson went into this Golgotha with hearty enthusiasm, gathering the fine white harvest of skulls spread before him, and throwing them in heaps like a boy gathering pumpkins. He brought nearly a hundred on board . . ."

In 1890 Nelson was commissioned by Merriam to collect specimens in California. Here he had occasion to stop for the night at a ranch and talk to the rancher about his need for an assistant. The rancher's son, it seemed, was interested in natural history and wanted to learn how to prepare animal skins. Thus began a long association between Nelson and Edward Alphonso Goldman (1873–1946). Goldman was then eighteen, Nelson "a black bearded young man of 36." The two were soon sent to Mexico by Merriam, the first of many trips there over a period of nearly fifteen years.

As chief of the U.S. Biological Survey from 1916 to 1927, Nelson promoted not only the Alaska Game Law but also the Migratory Bird Conservation Act, which provided for the first federally maintained refuges. Nelson's contributions are commemorated in the names of at least eight subspecies of birds and more than twenty species and subspecies of mammals, mostly from Mexico. One of them was, of course, *Neotoma nelsoni*, named by Goldman. Merriam named the genus of the Mexican diminutive wood rat *Nelsonia*. In addition, two species of reptiles, five species of fish, and two butterflies were named for Nelson. Nelson Island, along the coast of Alaska, stands as a tribute to his research in that state; it is separated from the mainland, in part, by Baird Strait, named, of course, for Spencer Baird.

Edward Goldman remained associated with the Biological Survey until his retirement in 1928. (In 1927 the Survey was combined with the Bureau of Fisheries to form the U.S. Fish and Wildlife Service.) Goldman was fluent in Spanish and well known in Mexico. He loved to tell stories of his adventures there. On one occasion he was attacked and robbed of his rifle, traps, and other possessions and left for dead. He eventually staggered to the nearest town, but he bore a scar on his temple for the rest of his life.

Goldman, too, had several birds and mammals named for him, including Goldman's spiny pocket mouse, Goldman's wood rat, and several others. Goldman Peak stands five thousand feet high in Baja California, named by Nelson. Goldman reciprocated by naming his oldest son Nelson. A closer intertwining of two lives would be hard to imagine.

Nelson's Larkspur

Larkspurs are among the showiest of plants, their generous blue blossoms borne on stalks well above most of the foliage. They are popular as garden plants, but the many wild species are not at all popular with ranchers, since they are among the most poisonous of plants. When cattle eat the fresh spring foliage in quantity they become violently ill, and there are records of hundreds of cattle dying in certain areas. Tincture of larkspur was once used externally to control head lice in humans. The poison is an alkaloid called delphinine. The name is based on the generic name *Delphinium*, which in turn is based on the Latin word for dolphin, *delphinus*. It was Linnaeus who named the genus, apparently because he saw the image of a dolphin in the flowers. The word *larkspur* is more appropriately descriptive of the flowers, the splayed petals of which suggest the forward toes of a lark, the backward-directed spur the long hind toe.

Nelson's larkspur, *Delphinium nelsonii*, brightens pastures and hillsides in parts of the West in spring. It was named for Aven Nelson (1859–1952), who came to the University of Wyoming as a professor in the year of its founding, 1886, and later served for several years as its president. Nelson had been hired to teach English, but by accident two professors of English had been hired and no one to teach biology. Nelson was assigned the task, and despite his initial weakness in the subject, he went on to a long and successful career as a botanist and founder of the Rocky Mountain Herbarium in Laramie. He and his students collected plants in Wyoming and adjacent states, and in 1909 he published a greatly expanded edition of John Coulter's *Manual of the Botany of the Rocky Mountain Region*. Nelson described many new species of plants, perhaps the most spectacular of which was giant angelica (*Angelica ampla*), a member of the parsley family that grows head-high on hollow stalks often an inch in diameter. Nelson's relationship with the eccentric botanist-preacher Edward Lee Greene was often rocky, but it was Greene who named Nelson's larkspur.

Two years after Nelson's wife died, in 1929, he married a graduate assistant, Ruth Ashton. Although he was then seventy-two years old, and she thirty-five, the two often traveled together to collect plants in out-of-the-way places, even Mount McKinley in Alaska. Ruth Ashton Nelson (1896–1987) published very useful, well-illustrated guides to the wildflowers of Rocky Mountain and Zion national parks. Aven Nelson died at ninety-three, Ruth at ninety-one. Clearly there is something to be said for a life of botanizing under bright western skies.

Ord's Kangaroo Rat

*I*t is rather too bad that so many rodents are labeled "mice" or "rats," even though they are remarkably diverse in appearance and in behavior. Kangaroo rats belong to quite a different group from the wood rats and deer mice. Their most distinctive feature is their external, fur-lined cheek pouches, which open forward on each side of their mouth. These are elegant little animals, with white feet and bellies, white hip stripes, and a long tail with a tuft of long hairs at the tip. The small front legs are used for manipulating seeds and hardly at all for walking. The large hind limbs enable them to leap about like diminutive kangaroos, often changing direction abruptly as they flee from a predator. They are well adapted for detecting the approach of a predator, too, since their relatively large heads contain an enlarged middle ear that is effective in picking up the low frequency sounds made by snakes and owls.

Kangaroo rats are best known to people who are themselves "desert rats," for they are usually seen by persons camping or traveling at night in deserts of the West. Ann Zwinger, in her book *The Mysterious Lands*, speaks of kangaroo rats as "quintessential desert rodents," since they live without imbibing water. They are able to create metabolic water and to pro-

Ord's Kangaroo Rat

duce hard, dry droppings; the urea they excrete is twenty to thirty times as concentrated as that in their blood. Their reproductive cycle is, however, geared to rains, since the moisture in fresh vegetation allows the female to produce milk, and the new plants will soon produce seeds enough to supply the young.

Seeds are gathered in the cheek pouches during nightly forays and stored in compartments in their complex, underground nests. Campers, sitting at a campfire in the desert, sometimes see the animals entering the circle of light, and if they approach slowly they can sometimes catch one by hand. Kangaroo rats are gentle creatures and will live for many months in captivity if provided with seeds, a place to hide, and soil with which to dust themselves. Since they are nocturnal, they make good pets only for people who themselves have nocturnal tendencies.

Although kangaroo rats seem like very specialized little animals, they actually have a long lineage. A fossil species has been found in rocks of northeastern Nebraska, dating from ten million years ago and accompanied by bones of two kinds of horses that are long since extinct. It has all the usual kangaroo rat features and even had a burrow containing a cache of hackberry seeds.

Ord's kangaroo rat (*Dipodomys ordii*) is the most widely distributed of the sixteen contemporary species of its genus in the United States, occurring from southern Canada to northern Mexico and from the central Great Plains to Nevada and eastern Oregon. A body only about four inches long is followed by a five-inch tail. On top the body is light brown, blending well with the sandy soils in which the animals thrive. There is a black streak on each side of the nose and a white stripe along each side of the blackish tail.

The diet of this and several other small desert-inhabiting rodents consists mainly of the seeds of grasses, although seeds of other plants are sometimes taken. In some sand dune areas several of these rodents occur together. Studies have shown that, to a degree, they avoid direct competition for food by sharing the seed supply. Ord's kangaroo rat, a medium-sized species, mostly takes seeds 2.4 to 4.7 millimeters in size, which are best suited for cramming into the cheek pouches. Larger species, such as the desert kangaroo rat, generally harvest larger seeds, while tiny species such as the little pocket mouse and the pale kangaroo mouse take very small seeds; the smaller species also often climb into bushes for seeds, while the larger species do not.

Seeds are cached in chambers in the nest and are a source of food during cool or stormy weather. The amount of seeds in storage may be large, sometimes several pounds. Deep in the soil, the seeds may become moldy or rotten. There is evidence that kangaroo rats can detect levels of moldiness and will move their caches about to insure that they do not deteriorate. Molds produce by-products that in low concentrations are beneficial, producing vitamins and antibiotics, but many molds are toxic in high concentrations. O. J. Reichman has shown that bannertail kangaroo rats prefer slightly moldy seeds to those that are very moldy or completely free of mold. Evidently they have evolved tastes that allow them to benefit from the by-products of molds without experiencing their toxic effects, and they manipulate their supplies of grain accordingly.

Seeds and fruits that are in advanced stages of rot or moldiness are extremely distasteful or toxic to animals. For example, corn moldy with *Penicillium* can, if eaten in quantity, kill a horse in three to five days. Tropical rodents rarely cache grain; the rapid growth of mold might prove fatal to them. We tend to take the distastefulness and toxic qualities of deteriorating foodstuffs for granted. We know it is to our advantage, and that of animals, to avoid such food items, but it is interesting to turn the matter on its head and ask why it is advantageous for the fungi and bacteria that cause rot to produce toxins. These microorganisms rely on fruits and seeds to live out their lives, which may last only a few days, so it is in their interest that fruits and seeds *not* be eaten by an animal. Daniel Janzen of the University of Pennsylvania has approached the subject from this unusual point of view. "Fruits rot, seeds mold, and meat spoils [writes Janzen] because that is the way microbes compete with bigger organisms." It is difficult for humans to develop a rapport with bacteria, fungi, and other very simple, mostly microscopic organisms, but they do live lives that on their terms are as significant as any others.

George Ord (1781–1866), for whom the kangaroo rat was named, has been called a "closet naturalist." His only long field trip was one to Florida in 1818, with Thomas Say and two others as companions. Primarily he was a philologist and lexicographer; he compiled much of the information for the first edition of Webster's *Dictionary*. Ord inherited a profitable business from his father and was able to retire at forty-eight and devote the remainder of his long life to erudite matters. He became active in the American Philosophical Society and through the society came to know many of the naturalists of the day. In 1815 he wrote a chapter entitled "North American Zoology" for William Guthrie's *Geography*, the first review of its kind. It qualified him, in the minds of some, as "the father of North American zoology." At the time of its publication the Lewis and Clark Expedition had been back only a few years, and it fell to Ord to describe several of the mammals and birds they had collected. Thus it was Ord who described the grizzly bear, which he appropriately named *Ursus horribilis*, as well as several other birds and mammals: Bonaparte's gull, the Columbian sharp-tailed grouse, the pronghorn antelope, the black-tailed prairie dog, the western gray squirrel, the Columbian ground squirrel, and the bushy-tailed wood rat. A formidable list of western wildlife for a person who had never been west!

It was Samuel Woodhouse who named Ord's kangaroo rat, from specimens taken near El Paso, Texas. Since I'll have more to say about Woodhouse in a later chapter, this seems an opportunity to discuss Ord's relationship with pioneer ornithologist Alexander Wilson (1766–1813). Wilson, a Scottish poet who had landed in jail for libel, migrated to Pennsylvania when he was twenty-eight and became a schoolteacher. He was intrigued by American birds and was welcomed at the Bartram Gardens, where William Bartram, whose *Travels* had made him well known throughout the English-speaking world, gave him access to his library. Without any training in art, Wilson began to draw birds, and, through trial and error, he became proficient at rendering accurate, if rather stiff, portraits of birds. He began to travel widely in order to add birds to his portfolio, in his own words "a solitary exploring pilgrim." Van Wyck Brooks presents an appealing picture of his peregrinations in *The World of Washington Irving*:

> He plodded on foot or on horseback through horrid swamps and sluggish creeks. . . . His pockets were crammed with the skins of birds, and a Carolina paroquet was his sole compan-

ion . . . it perched on his shoulder, ate from his mouth and even responded to its name, and it always amused the Indians whom he passed on his way. To beguile his lonesome march he played Scottish airs on his flute. . . .

In 1806 Wilson wrote to Thomas Jefferson that he had over one hundred drawings completed for an *American Ornithology*, "and two plates in folio already engraved." Wilson was given the opportunity to describe some of the birds taken on the Lewis and Clark Expedition, and it was he who described Clark's nutcracker and several other birds in his *American Ornithology*. This many-volumed work contained not only his plates, hand-colored by artists whom he hired, but also a text rich in mostly accurate detail and told with a touch of poetry. Thomas Jefferson was one of the first subscribers.

The story of the meeting between Wilson and Audubon has been told many times. Traveling down the Ohio on a skiff he had named the *Ornithologist*, in 1810, Wilson stopped at Shippingport, near Louisville, and called at the store of Rozier and Audubon. The meeting of the two came as a not wholly pleasant surprise to both. Audubon later remembered Wilson's "long rather hooked nose, the keenness of his eyes and prominent cheekbones [which] stamped his countenance with a peculiar character." Audubon was, however, too poor to subscribe to Wilson's project, and he probably felt that Wilson's drawings were inferior to his own. He showed Wilson his paintings and Wilson asked to borrow some of them. The two hunted together for a time and took birds that Wilson had not seen before. Wilson was understandably discouraged to find that he was not the only person engaged in drawing American birds and disgruntled that a man of similar interests would not subscribe to his *American Ornithology*. But the legend of the Wilson-Audubon dispute was largely born in the mind of George Ord, and well after Wilson's death.

Ord had probably met Wilson at Bartram's Gardens about 1803. He greatly admired Wilson, fifteen years his senior, and the two sometimes hunted birds together. Wilson died in 1813 after contracting dysentery from swimming a river to collect a bird. He had finished eight volumes of *American Ornithology*. Ord undertook to finish volume nine and write volume ten from Wilson's notes. He also wrote a biographical sketch of Wilson and edited his journals. There is good reason to believe that Ord, "a small man of sour impulses" (in Peter Mattheissen's words), made changes in Wilson's text and in his journals, making it seem that Audubon had seriously snubbed Wilson during the meeting at Shippingport. Ord disliked Audubon personally, and he objected particularly to the landscapes, flowers, and trees that Audubon used as backgrounds to his bird portraits. He marshalled his forces in defense of Wilson, inciting help from the influential English naturalist Charles Waterton, whose *Wanderings in South America* was just then on everyone's bookshelf.

Audubon not only refuted Ord's account of the episode at Shippingport but went on to accuse Wilson of copying his drawing of a flycatcher. This led to a counterattack by Ord in which he charged Audubon with copying Wilson's drawings of a female red-winged blackbird and a Mississippi kite (a charge that may have been justified). Wilson himself, though in Henry Adams's words "a confirmed grumbler," had done little to bring about such a controversy, and he had now been dead several years. In the final reckoning, neither his reputation nor Audubon's suffered greatly from Ord's interference.

In his *American Ornithology*, Wilson described as new fifty-six species of American birds, about forty of which still stand as valid species. Some of our best-known birds are among them: the song sparrow, marsh wren, and whip-poor-will, for example. The genus *Wilsonia* was named by Charles Bonaparte, who produced a supplement to the *American Ornithology*; the genus includes several of our most attractive migrants: the Canada, hooded, and Wilson's warblers. Wilson's phalarope, Wilson's plover, and Wilson's storm-petrel also bear his name.

Elliott Coues wrote that Wilson had "genius and not much of anything else—very little learning, scarcely any money, not many friends and a paltry share of the world's regard while he lived." But, wrote his biographer Robert Cantwell, "the *American Ornithology* was so rare, strange and hauntingly beautiful that it transformed a work of science into a work of art. . . ."

Palmer Saltgrass

*O*ne of the most frightening of our current environmental
problems is that while human populations continue to
grow exponentially, much agricultural land is being lost through
increased salinization. Evaporation from irrigated soils causes
salt to accumulate in the upper few inches, and the reuse of irrigation water downstream produces a further accumulation of
salts. Salt buildup in the Tigris-Euphrates Valley of the Near
East centuries ago destroyed the world's first agricultural
region. Today salinization is a serious problem in parts of Asia
and Africa, in the lower Colorado River Valley of southern Arizona and California, and especially in adjacent parts of Mexico.
California's San Joaquin Valley and parts of the Great Basin are
also threatened. According to one estimate, four billion acres of
land, throughout the world, have become too saline for conventional agriculture. Most of this acreage is in third world countries, in many of which undernourishment is already a problem,
but here in the United States 200,000 to 300,000 acres are lost
each year to salinization.

There are ways of using our technology to combat these
problems, but these provide no more than delaying actions. It is
urgent that plants be found or bred that will provide food crops

when grown in salty soil. Plants that grow in such soils (halo-phytes) occur mainly along seacoasts and salt lakes. Using four-wheel-drive vehicles, aerial reconnaissance, and satellite photographs, researchers at the University of Arizona's Environmental Research Laboratory have been exploring coastlines in the search for potential new food plants that will thrive in saline soils. It was known that in 1889 Edward Palmer found the Cocopa Indians harvesting grain from a halophytic grass on the Colorado River estuary, at the upper tip of the Gulf of Cal-

Palmer Saltgrass

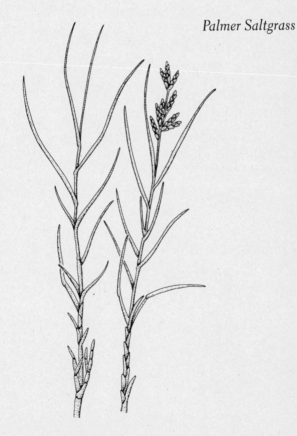

ifornia. Palmer sent specimens of the grass to the U.S. Department of Agriculture, where their botanist, George Vasey, gave them the name *Distichlis palmeri*. Then the plants were forgotten for many years, and it was thought that the species might be extinct. But University of Arizona researchers discovered several patches, and they collected seeds and planted them in experimental plots. Remarkably, the seeds are not salty, even though these plants grow on tidal flats that are periodically flooded with seawater.

The seeds of Palmer saltgrass bear much resemblance to wheat, and the Native Americans ground them into a flour used for making soups and breads. In fact they contain three times as much fiber as wheat and lack gluten, a constituent of wheat that produces allergies in some people. Furthermore, this is a perennial grass and can be harvested periodically without replanting. One disadvantage is that the spikelets shatter and discharge the seeds freely. However, the seeds float, and they were gathered by the Native Americans as they washed up in great numbers at high tide. Natural populations of the saltgrass do not have a high yield, but hybrid cultivars have now been developed that produce as much as 890 pounds per acre. In taste tests, muffins made from this cultivar (patented as Wild-Wheat) were found to be as acceptable as those made from wheat flour. Further research is being conducted on this and other halophytes, giving promise that some of the vast acreages of salinized soils throughout the world may someday be put to use for growing food crops.

It is worth noting that *Distichlis palmeri* had attracted little attention since its discovery a century ago. Yet suddenly it is in the limelight and may even, eventually, prevent starvation among millions now living where their soils can no longer grow conventional crops. Nowadays one often hears that because there are so many species in the world, many still undescribed (perhaps more than half), we should ignore the past accumulation of naturalists and from now on study only "important" species. But how do we know which may someday prove important?

Edward Palmer (1831–1911) is not nearly as well known as he deserves to be, perhaps because he was in no sense a professional but made his living wandering the wilds and collecting specimens of all kinds, supporting himself by selling specimens and by taking temporary jobs from time to time. He was much interested in Native American culture, and his article *Products of the North American Indians* (1871) has been reprinted many times. Among his other accomplishments, he was the first to call attention to a small weevil which was so destructive to cotton in parts of Mexico that growers were giving up planting cotton. Palmer sent specimens to the U.S. Secretary of Agriculture, who passed them on to the federal entomologists. It turned out to be an obscure weevil described in 1843 by a Swedish entomologist— another seemingly unimportant organism that was suddenly important. In fact within fifty years the cotton boll weevil had spread throughout the cotton belt of the United States, causing an estimated $5 billion in damage—at a time when the dollar was worth a good deal more than it is now.

Palmer was born in England but moved to America when he was eighteen. Settling for a time in Cleveland, he met Jared Kirtland, who encouraged him to begin collecting natural history specimens. In 1852 he applied for a position as a medical steward on the ship *Water Witch*, on its way to Paraguay. He was accepted and managed to find time to collect biological materials for the Smithsonian Institution at various stops along the way. On his return he tried to obtain a position at the Smithsonian, but failing that he enrolled in a few lectures on medicine— enough to qualify him as a "doctor" in the mid–nineteenth century. The next few years found him in the West, and when the Civil War broke out he served in various Union army hospitals in the West. When he could he collected plants and animals and sent them to George Engelmann and Spencer Baird. Posted to Arizona in 1865, he met Elliott Coues, and the two collected together until both were posted elsewhere.

Palmer's eagerness as a collector can be appreciated from the fact that following an army raid on an Apache village, he collected human remains from among the dead; an Apache child was to him only a specimen. Earlier, in Paraguay, he had tried to measure the size of the imposing breasts of Indian women. Throughout his life he collected specimens, including pottery and chips from buildings and monuments, with little awareness that he sometimes overstepped the bounds of rationality.

After the Civil War, Palmer roamed the West, collecting for Baird, Engelmann, Parry, Torrey, Gray, and others. After a collecting trip he would often advertise his wares at so much per set of duplicate specimens. For example, after a trip to Mexico in 1879–1880, he distributed sixteen lots of duplicate plants in part as follows:

1,441 specimens to W. Canby (N.Y. Botanical Garden)	$139
1,433 specimens to Kew Gardens, London	$138
1,288 specimens to George Vasey (U.S. Nat. Herbarium)	$124
952 specimens to J. Smith (private collector)	$ 80

The funds from these sales were then used to finance further collecting trips. For a time Palmer was employed by the Bureau of Ethnology, for another period of several years by the Department of Agriculture. Without doubt his many forays into Mexico, some with Charles Parry, yielded the most valuable collections of unusual plants.

Palmer was of small stature, continually complaining about his health, although at seventy-five he was still traveling by horseback in the mountains of Mexico. He seemed to have little sense of humor and constantly complained of being mistreated by others. From time to time he feuded with Coues, Parry, and others. But he was a collector like no one else and is said to have taken over 100,000 specimens of plants, as well as many birds, mammals, insects, and Native American artifacts. His few publications are much valued, as they dealt, in most cases, with tribes now gone or much changed. Asa Gray named a genus of plants for him, *Palmerella*, and many species, mostly Mexican, bear his name. Among species occurring in the United States may be mentioned Palmer's agave, named by Engelmann, Palmer's penstemon, named by Gray, Palmer's chipmunk, named by Merriam, and a subspecies of the curve-billed thrasher, named by Coues.

George Vasey (1822–1893), who described Palmer saltgrass, was at the time in charge of the U.S. National Herbarium, which had been transferred from the Smithsonian to the Department of Agriculture in 1868. Like Palmer, Vasey was born in England but moved to America when quite young. At his home in central New York he began collecting wildflowers, and a local doctor put him in touch with John Torrey and Asa Gray. He later obtained a medical degree and established a practice in Illinois. In 1868 his friend and neighbor, John Wesley Powell, invited him to join his Rocky Mountain Scientific Exploring Expedition as botanist. Vasey left the expedition before the historic descent of the Grand Canyon of the Colorado, but some months later, deep in Marble Canyon, Powell found a grotto filled with ferns and wildflowers and named it Vasey's Paradise. It is still marked on some maps.

Vasey's appointment to the National Herbarium a few years later was questioned by some, as he had published little up to that time. But he set to work organizing the large amount of material that had accumulated and within a few years began to publish voluminously. He soon became the leading authority on the grasses of North and Middle America. He named many species, mostly from Mexico, and of course had a few named for him. He depended upon Palmer and others to supply specimens, as he did not travel much after his trip with Powell. He is said to have been gentle and kindly, in contrast to Palmer, who had a more vituperative personality.

Queen Alexandra's Sulphur Butterfly

*T*here is no more delightful experience than to come upon a drying puddle crowded with yellow butterflies, their long tongues probing for moisture and minerals; then to disturb them just enough to cause them to fly up and surround one with a cloud of fragile slips of gold. Or sulphur—for these are called sulphur butterflies. Or butter—doubtless it was such as these that long ago inspired the word butterfly. More than fifteen kinds of sulphur butterflies grace our fields and meadows, all of them rather similar in appearance. In the West, one of the most prevalent is Queen Alexandra's sulphur, *Colias alexandra*, distinguished from closely related species by the somewhat pointed outer angles of the front wings and the narrow black border of the wings of the males. In the females, the dark border is broader but fainter and more irregular. In both sexes, the undersides of the wings are duller, flecked with greenish scales. The larvae are green, with a white stripe along their sides, blending well with the plants on which they feed: milk vetch, loco weed, and golden banner. They overwinter partly grown, then feed rapidly in the spring, make a chrysalid, and emerge as butterflies in June.

It is easy to take these lovely bits of winged sunlight for granted, as we do so many things. But there is more to their lives than we might imagine. Often pairs are seen spiraling upward, sometimes as much as sixty feet into the sky. Invariably one is found to be a male, the other a female that has recently mated. This is an avoidance flight on the part of the female; as she ascends, the male eventually drops out, leaving the female free to find a place to lay her eggs. (Most of the research has been done on the common sulphur and the alfalfa butterflies, which are closely related to Queen Alexandra's and probably behave very similarly.)

Virgin females, in contrast, will sometimes chase males, since during copulation males transfer to the females a packet of sperms and protein (called a spermatophore). Females require this to produce a full complement of viable eggs. When

Queen Alexandra's Sulphur Butterfly

a female has used up the contents of the spermatophore, she is once again receptive to a male. But not every male will be equally able to produce a large spermatophore; smaller males, and those that have recently mated, have less to offer. Ronald Rutowski of Arizona State University found that males (of a related species, the checkered white) courted females more vigorously the larger the spermatophores they bore: in this sense they were "honest salesmen."

Males are able to produce several successive spermatophores, and females are often able to mate and lay eggs several times. It is the males with the greatest longevity and flight capacity that father the most offspring. Investigators at the Rocky Mountain Biological Laboratory have shown that the enzymes that control flight energetics are located on specific genes. Males that succeed in mating several times with older, more "choosy" females are in effect transmitting genes for superior flight capacity—a capacity that is advantageous to both sexes in many aspects of their lives. It has often been asked whether choice of a mate (frequently based on trivial aspects of color or courtship behavior) can result in the production of truly superior offspring. In other words, how can an individual select a mate that will provide the best genes suitable for his or her particular life-style? Sulphur butterflies seem to have provided an answer.

Colors are all-important in the courtship and mating of butterflies, and since the various species of sulphurs are all colored much the same, one wonders how they sort themselves out in nature. In fact, in agricultural areas common sulphurs and alfalfa butterflies do hybridize to some extent (the larvae of both feed on alfalfa), although generally they remain distinct species. It has been known for many years that butterflies, bees, and some other insects do not see colors as we do—ultraviolet (UV) is part of their visible spectrum. By equipping a television camera with a UV-transmitting lens and directing it at natural objects, it is possible to see these objects much as insects see them. Many flowers have patterns of UV that direct bees and butterflies to the nectaries. Many butterflies, too, have patterns of scales that reflect UV. The reproductive isolation of the species of sulphurs can be explained, at least in part, on the basis of these UV patterns.

Males of the alfalfa butterfly (*Colias eurytheme*) have been found to have a brilliant pattern of UV reflection on the upper surface of the wings; such a pattern is absent in males of the common sulphur (*Colias philodice*) and in the females of both species. Male alfalfa butterflies respond to dummies made of yellow paper lacking UV, but if UV is added they are inhibited, since this signals to them that the dummy is a male. Transvestite alfalfa butterflies, made by transposing female wings onto male bodies, attract males, indicating that they respond to the wings and not the body. Females reject males of their own species when the wings are treated so that they lack UV reflection. When the males fly about, the UV pattern flickers on and off, since it is on the upper surface only; clearly it is a releaser of sexual behavior.

Colias alexandra is a highly variable species. Males in the southern Rocky Mountains are yellow but have a large patch of UV reflection on the hind wing. Farther west and north the males are suffused with orange by ordinary light and have streaks of UV on the fore wings as well as a patch on the hind wings. In the Dakotas, north to Alaska, males are orange and have UV over much of the fore and hind wings. Lepidopterists recognize several subspecies, perhaps as many as ten, based on these features and others. If the UV patterns play a major role as mating signals, it seems likely that the various subspecies interbreed little if at all.

Some of the important research on the reproductive behavior of sulphur butterflies was done by Robert Silberglied, a professor at Harvard and a staff scientist at the Smithsonian

Tropical Research Institute. Tragically, the promising career of this brilliant and versatile young naturalist was cut short when he was killed in an airplane crash in 1982 at age thirty-six.

Colias alexandra was described by William Henry Edwards (1822–1909) in 1863, the year that Alexandra, daughter of the king of Denmark, married Albert Edward, son of Queen Victoria and her consort, Prince Albert. Alexandra was the "fairy princess," so beautiful and charming that the world was agog. (Presumably the butterfly was called Princess Alexandra's sulphur until 1901, for it was not until then that Edward became king and Alexandra queen.) Besides the sulphur, Queen Alexandra's birdwing, the largest butterfly in the world, with a wingspan of eleven inches, was named for her. It is a native of New Guinea and is now very rare as a result of deforestation and the persistence of collectors. It was named by the wealthy and eccentric Lord Walter Rothschild, who knew Alexandra well. In fact, when he drove his carriage into the forecourt of Buckingham Palace, pulled by zebras, the princess tried to pet one of the zebras. This alarmed Lord Rothschild, whose zebras had never been completely tamed. Rothschild built his own zoo and museum, the latter containing over two million specimens of butterflies and moths gathered from all over the world.

Alexandra devoted her life to her five children and to maintaining the pomp and circumstance that surround the British monarchy, while Edward VII "continued to consume his women with the same energy with which he consumed his mountainous meals" (in the words of a biographer). Neither was a naturalist in any sense of the word, and Edward was a dedicated antinaturalist. On his estate he and his guests slaughtered thousands of deer and other animals each year. In Egypt he prided himself on having killed a crocodile as well as hoopoes, spoonbills, flamingos, cranes, and other wildlife. From India in 1876 he wrote to Prince George (later King George V): "I have had great tiger shooting. The day before yesterday I killed six. . . . Today I killed a tigress and she had a little cub with her." This has little to do with our story except as a demonstration of the other side of the coin—quite a different way of taking pleasure in nature.

William Henry Edwards, a great-great-grandson of New England preacher and philosopher Jonathan Edwards, was brought up in the Catskills of New York, where he began collecting butterflies when quite young. When he was twenty-four he traveled to Brazil to collect natural history specimens. His book *Voyage up the River Amazon* was read by English naturalists Henry Walter Bates and Alfred Russell Wallace and inspired them to undertake their epic journey a few years later. Edwards was trained in law and involved in various business enterprises, including railroads and coal mines, but butterflies remained his passion. His three-volume *Butterflies of North America* (1868–1897) was beautifully illustrated by carefully trained artists and was a milestone in the study of these insects. Edwards was distressed that no one had studied the larvae of butterflies since the time of John Abbot, so he undertook to collect eggs from captured females and rear the caterpillars on suitable host plants. Completion of his magnum opus threatened to exhaust his personal finances, but fortunately another well-known lepidopterist, W. J. Holland, director of the Carnegie Museum in Pittsburgh, agreed to meet many of the expenses of the third volume with the understanding that Edwards would deposit his collection at the museum, which he did. Holland is best known for his two popular volumes *The Butterfly Book* and *The Moth Book*.

Edwards carried on an extensive correspondence with Spencer Baird at the Smithsonian and at times helped to support collecting expeditions. For example, he paid Robert Kennicott $50 a year if he would collect Lepidoptera while in the Far North, and he paid John Xántus $100 to collect butterflies in Baja California. He was not always satisfied with the results.

The specimens from Clarence King's travels in California he found "worthless, all common species & all eaten by insects."

When the expedition of George Wheeler headed west, Edwards helped a young man, Theodore L. Mead, to join the group. Mead wrote the report on Lepidoptera for the publications of the Wheeler Expedition, and in 1882 he married Edwards's daughter Edith, for whom he named a butterfly, Edith's copper. Edwards himself did not make a collecting trip west until 1894, when he was in his seventies. Between them, Edwards and his son-in-law named a great many of the West's butterflies, and both had several named for them. Edwards's fritillary (*Speyeria edwardsi*) is one of the larger and more brilliantly colored butterflies of the Rockies; Mead's alpine is a dusky butterfly of meadows above timberline. Edwards was described by a friend as "kind, open-hearted, cheery, and courteous, free from pride and ostentation"—surely suitable qualities for a person who spent his life with butterflies.

Queen Anne's Lace

The story is told that Queen Anne of England once challenged her ladies-in-waiting to make an embroidery as exquisite as the umbel of flowers of this common roadside plant. Whether they succeeded is unknown, but the common name has stuck, a better name than the bland one provided by Linnaeus, *Daucus carota* (*daucus* is Greek for carrot, *carota* Latin). Queen Anne's lace (or wild carrot) is the ancestor of the carrots we grow in our gardens; it was brought from Europe to America by early settlers. Despite its filigreed blossoms, so attractive to bees and butterflies, it is often regarded as a useless weed. The slightly unpleasant odor evidently has suggested medicinal uses; the

Romans used the plant as an aphrodisiac, and in early England it was recommended for dropsy and for flatulence.

Anne, who reigned from 1702 to 1714, was evidently a satisfactory queen, though she was corpulent and had to be moved about in a chair or by pulleys. She was ill much of her life, and none of her seventeen children lived for more than a few years. But she loved her gardens and, if the stories are true, had a special affection for the plant that bears her name.

When the *New York Times*, in one of its pastoral moods, published an account of the naming of the plant for Queen Anne of England, it was challenged by one of its readers, who suggested that it may have been named much earlier for Saint Anne, sometimes referred to as "the queen of heaven." Anne was the mother of Mary and the grandmother of Jesus as well as the patron saint of lacemakers. According to this correspondent, Martin Luther attacked the worship of Saint Anne, perhaps resulting in the invention of a Protestant explanation of the name. Who can say?

Rafinesque's Big-eared Bat

*B*ats have long been the inspiration of fear and supersti-
tion. Dark forms of the night, they were rumored to
become entangled in women's hair, to carry mysterious diseases, to foretell disaster and death. The association of bats
with vampires and witches was reinforced by the publication of
Bram Stoker's novel *Dracula* in 1897 and the several movies
that have since been made from it. Stoker was probably influ-
enced by tales of blood-feeding bats brought back by travelers
in the American tropics. The Maya Indians, who lived within
the range of these bats, had a bat god, Zotz, who lived in the
underworld and also provided the symbol for one of the months
of the year in the Mayan calendar. But the association of bats
with death has long prevailed in the Old World, too, though
there are no blood-feeding bats there. Bats were sacred to
Persephone, wife of Hades, ruler of the underworld of Greek
legend, and throughout Eurasia and Africa bats have been associated with ghouls and witches. Considering the fact that the
vast majority of mammals are creatures of the night, it is hard to
understand why bats in particular have been suspected of evil-
doings. Perhaps it is in part because, being flying creatures, they

contrast so sharply with birds, lacking the bright colors and melodious songs that do so much to brighten the sunlit hours.

There is good reason to avoid the attacks of the vampire bats of the American tropics. It has been estimated that a colony of about one hundred might, in a year, consume the amount of blood in twenty-five cows. Vampire bats transmit rabies and other diseases to livestock and humans, and the insectivorous bats of temperate North America sometimes also transmit rabies if handled carelessly. In the last thirty years, about eight deaths have been reported in the United States and Canada resulting from the bites of bats. This is a much lower death rate than that from bee stings or dog bites, and of course the figure

Rafinesque's Big-eared Bat

is infinitesimal compared to the tens of thousands of deaths from automobile accidents.

By and large, bats are beneficial animals, since the majority feed on insects that they snag during their nightly flights. One Mexican free-tailed bat—they emerge by the millions from Carlsbad Caverns every evening—has been found to take at least a gram of insects each evening. According to one estimate, the bats in this colony may take 6,700 tons of insects in the course of a summer. The insects include mosquitoes and a variety of moths and beetles that are pests of agriculture. Bats have been used in medical research, since the wing membranes are so thin it is possible to observe directly the effects of drugs on blood vessels. John E. Hill and James D. Smith in their book *Bats: A Natural History* (the source of much of this information) discuss a proposed use of bats during World War II. Mexican free-tailed bats were equipped with small incendiary bombs and were to be dropped over enemy territory. Unfortunately some escaped and set fire to several buildings in the Southwest. They were never used in combat. Finally, some of the large fruit bats of East Asia, the East Indies, and Australia have long been used as food and are considered a delicacy by the natives.

The diversity of bats is almost beyond belief. Giant fruit bats ("flying foxes") may have a wingspan of nearly six feet, while the smallest bat has a wingspan of no more than five inches. The diversity in form and color is apparent from a mere listing of names: the leafnose bat, the hognose bat, the mastiff bat, the red bat, the spotted bat, the hoary bat. These are North American species; tropical species include not only the fruit bats and vampires, but also the hammerhead bat, the proboscis bat, the fish-eating bat, the fringe-lipped or frog-eating bat, the sucker-footed bat, the club-footed bat, and so on: enough to drive one batty. Altogether there are about 950 species in the world, about 40 in the United States and Canada.

Rafinesque's big-eared bat (*Plecotus rafinesquii*) is one of the more distinctive species, with ears more than an inch long, even though the body measures only four inches, the wingspan about ten inches. When they rest by day, they relax their ears by withdrawing blood from them and use them to wrap themselves. Then when ready to take flight, they pump blood into their ears so that they become fully erect. These are attractive bats, with bicolored fur and white underparts, though it is better not to examine them too closely; two large humps just above the nose give them a ghoulish expression. This is a species of the southeastern United States, ranging north to the Ohio Valley. During the day the bats rest in hollow trees, behind loose bark, in abandoned buildings, or less often in caves. In summer, females form dense colonies where they nurse their young, leaving them alone at night while they are out hunting insects. Males are usually solitary at this time, but later in the season they join the females for mating. During the winter the bats remain deep within caves or mineshafts, where they often move about restlessly. In the spring each female gives birth to a single batlet.

Insectivorous bats tend to have large ears, which form a necessary adjunct to their ability to pursue and capture prey by echolocation, that is, by producing ultrasonic pulses and responding to their echoes. (In contrast, fruit bats usually have small ears, as they have less use for sonar.) The ears of Rafinesque's big-eared bat are unusually large and quite mobile. When a resting bat is approached, he begins to wave his ears, perhaps to better keep track of the invader. There is some evidence that bats with very large ears may use them not only to pick up high-frequency sounds bounced back from objects but also the low-frequency, low-amplitude sounds made by flying insects.

It has been known for a long time that bats depend upon their ears more than their eyes for navigation and prey capture. An Italian physicist, Lazzaro Spallanzani, in the late 1700s found that blinded bats readily avoid objects in their path and catch as many insects as normal bats. But when he plugged their ears, they blundered into objects in their path. It was not until 1941 that it was clearly demonstrated that bats can "see with their ears" by producing ultrasonic pulses from their larynx that are bounced back from objects and are collected by their ears. These sounds are largely inaudible to us, but if we could hear them they would seem very loud; it is said that the calls of a little brown bat a few feet away might sound like a pneumatic drill. By means of its sonar, a little brown bat can detect a wire only one-eighth inch in diameter at a distance of seven feet. Why ultrasound rather than sounds of lower frequency? For one thing, bats can thereby eliminate interference from background noises, such as wind, movements of animals, songs of insects, and so forth. For another, the short wavelength of high frequency sounds is better for detecting the position and shape of small objects such as insects. By means of echolocation, bats are able to determine the presence of a prey insect, estimate its distance and direction, and finally capture it if it is found to be something worth eating.

Donald Griffin of Rockefeller University, while still a student at Harvard, was the first to demonstrate clearly the importance of sonar in the lives of bats. I well remember hearing Griffin lecture to a startled audience, and I remember the impact made by his 1958 book *Listening in the Dark*. A few years later Kenneth Roeder of Tufts University showed that some moths are able to hear the ultrasound of bats by means of simple ears on the sides of their bodies. At the approach of a bat they drop or fly in a devious pattern, often avoiding capture. Certain moths can even produce high-pitched sounds that evidently "jam" the sonar of the bats. This is a field of continuing surprises.

Constantine Samuel Rafinesque (1783–1840) found and described the big-eared bat in 1818. He called it *Vespertilio megalotis* (*vespertilio* is the Latin word for bat, *megalotis* Greek for big-eared). However, this name had been used earlier for another species of bat, so it had to be renamed. This task fell to French naturalist René Primevère Lesson (1794–1849). He moved the species to the genus *Plecotus* and renamed it *rafinesquii.* Lesson is best known as the surgeon and naturalist on the round-the-world voyage of the *Coquille* in 1822–1825. As a result of his experiences in remote places he became enthusiastic about birds, and he described species from many parts of the world. Hummingbirds were his special favorites, and he named the blue-throated hummingbird of Mexico and our deep Southwest for his wife, calling it *Lampornis clemenciae.* His publications included not only a *Manuel d'Ornithologie* but also a *Manuel de Mammalogie.*

Rafinesque was the most colorful and the most exasperating of all nineteenth century naturalists. He was born in Constantinople of a French father and a Greek mother of German parentage. Much of his childhood was spent in Marseilles. "My botanical walks near Marseilles gave me much pleasure," he wrote in his autobiography, *A Life of Travels.* "I began the study of the Fishes and Birds, I drew them, and collected Shells and Crabs." When he was nineteen he and his brother were sent to Pennsylvania, where he "became gradually acquainted with all the Botanists, Naturalists and Amateurs of that period." He traveled about the eastern states, visited the Bartram Gardens, and everywhere collected plants, birds, and reptiles, most of them, according to him, previously unknown species. In 1805 he was offered a position in Sicily, where he remained for ten years, often botanizing with William Swainson, who was also there at that time.

On his return to America in 1815, he was shipwrecked not far out of New York. As he wrote in his *Life of Travels,* "I had lost everything, my fortune, my share of the cargo, my collections and labors of 20 years past, my books, my manuscripts, my drawings, even my clothes. . . ." When his wife in Sicily heard of his shipwreck, she ran off with an actor. These were but the first of his many misfortunes, none of which, however, ever dampened his enormous self-esteem.

Rafinesque made it to New York, where he was accepted as a member of the newly formed Lyceum of Natural History. But he soon became restless, and in 1818 he set off on the first of several trips across the Appalachians. In Henderson, Kentucky, he looked up Audubon, who described the visit in an essay titled "The Eccentric Naturalist" in his (somewhat fictionalized) *Ornithological Biography*.

He pulled off his shoes, drew his stocking so as to cover the holes about his heels telling us all the while in the gayest imaginable mood that he had walked a great distance. His agreeable conversation made us forget his singular appearance. A long loose coat of yellow nankeen cloth—stained all over with the juice of plants. . . . His beard was long, his lank black hair hung loosely over his shoulders. . . . I laid my portfolios open before him. He turned to the drawing of a plant quite new to him, inspected it closely, shook his head and told me no such plant existed in nature. I told my guest the plant was common in the immediate neighborhood. He importuned: "Let us go now." We reached the river bank and I pointed to the plant. I turned to Rafinesque and thought he had gone mad. He began plucking the plants one after the other, danced, hugged me, told me exultingly that he had not had now merely a new species, but a new genus.

Rafinesque spent the night with Audubon, and Audubon heard him late at night running about his room and banging the wall. He had found Audubon's violin and was using the bow to knock down the bats in his room, not because they bothered him, but because they might be a "new species." Audubon decided to test his gullibility by sketching several fishes he claimed to have seen. Rafinesque copied these into his notebook and later described and named them in his monograph on the fishes of the Ohio River. One of these, "the devil-jack diamond fish," was said to be four to ten feet long and to have scales so hard they were bulletproof and produced a spark when struck with steel. The scientific name he proposed (needless to say now in the zoological wastebasket) was *Litholepis admantinus* (Greek for hard stone-scale).

Rafinesque applied for a position at the newly founded Transylvania University in Lexington, Kentucky. He spent seven years there as professor of natural history. While there he published over two hundred articles on a wide variety of subjects. His lectures were popular, illustrated as they were with specimens he had collected nearby.

One student wrote of "his queer French accent" and his careless garb. "As he proceeded with a lecture and warmed up to his subject, he became excited, threw off his coat, his vest worked up to make room for the surging bulk of flesh and the white shirt which sought an escape, and heedless alike of his personal appearance and the amusement he furnished, was oblivious to everything but his subject."

Following his dismissal from Transylvania after a quarrel with its president, he moved to Philadelphia, where he continued to produce volumes of publications, including an epic poem about nature that he modestly compared to Milton's *Paradise Lost*. By this time some scientific journals, including the prestigious *American Journal of Science*, began declining his articles, and some of his fellow naturalists were becoming disturbed by his rashness in bursting into print with so many short descriptions of novelties, without careful consideration of the work of others. His "passion for establishing new genera and species," wrote Asa Gray, "appears to have become a complete monomania." "Rafinesque certainly deserves to be ridiculed, wrote John Torrey. "His vanity is absolutely intolerable."

Yet there was a streak of genius in Rafinesque. William Dall wrote that he had "a mind of unusual acumen, brilliancy, and activity . . . always clouded by a certain incoherency, due to his highly excitable and versatile temperament. He possessed talents which, properly regulated, would have carried him to the front rank of scientific workers." Louis Agassiz wrote of him: "From what I can learn of Rafinesque, I am satisfied that he was a better man than he appeared. His misfortune was his prurient desire for novelties and his rashness in publishing them, and yet both in Europe and America he has anticipated most of his contemporaries in the discovery of new genera and species."

In his later years, Rafinesque came to believe that new species and genera were continually being produced through changes in existing plants and animals. Such ideas, expressed well before the publication of Darwin's *Origin of Species*, further alienated him from his colleagues, to whom such ideas were anathema. Rafinesque died alone of stomach cancer in a Philadelphia garret at the age of fifty-seven, surrounded by his specimens and copies of his over nine hundred publications. His personal effects were sold at auction, and his body was locked in his room by his landlord, who hoped to sell it to a medical school. But a friend broke in through a window, lowered the body on a rope, and saw to it that it was properly buried.

How many of the names Rafinesque applied to species and genera are still in use? A great many, including both plants and animals. Several genera and more than thirty species of fishes bear the names he gave them, and many are well known, such as the bluegill, the white crappie, the black bullhead, and the rock sturgeon. Besides the bat, the genus of the desert chicory of the Southwest, *Rafinesquia*, was named for him (by Thomas Nuttall).

Rafinesque left behind an expression of his own philosophy that will do for that of many a naturalist.

A life of travels and exertions has its pleasures and its pains, its sudden delights and deep joys mixt with dangers, trials, difficulties, and troubles. . . . Mosquitoes and flies often annoy you or suck your blood. Ants crawl on you whenever you rest on the ground, wasps will assail you like furies. . . . You may be overtaken by a storm, the trees fall around you, the thunder roars and strikes before you. You may fall sick on the road and become helpless. Yet many fair days and fair roads are met with, a clear sky or a bracing breeze inspires delight and ease, you breathe the pure air of the country, every rill and brook offers a draught of limpid fluid.

Every step taken into the fields, groves, and hills appears to afford new enjoyments. Here is an old acquaintance seen again; there is a novelty, a rare plant, perhaps a new one! greets your view: you hasten to pluck it, examine it, admire, and put it in your book. Then you walk on thinking what it might be, or may be made by you hereafter. You feel an exultation, you are a conquerer, you have made a conquest over Nature, you are going to add a new object, or a page to science.

Richardson's Ground Squirrel

While Rafinesque struggled with mosquitoes and storms for the sake of finding new bats, fishes, and plants in the Ohio Valley, others were undergoing much more severe tortures in the Far North and making discoveries in a wholly new fauna and flora. John Richardson (1787–1865), physician and naturalist on John Franklin's first two expeditions into the Arctic, made important collections in central Canada and later published his

Fauna Boreali-Americana, the first comprehensive study of the fauna of arctic and subarctic North America. Six subspecies of mammals are named for Richardson on the basis of his collecting, and two species: the water vole, *Microtus richardsoni*, and a ground squirrel, *Spermophilus richardsoni*. Richardson's ground squirrels are not glamorous animals; they lack the bushy tails of tree squirrels and the decorative stripes of chipmunks. But they are abundant enough in parts of the West to insist on our attention. They communicate with both sounds and odors; besides a warning "chirp," they have glands that produce scents that both attract and repel other members of their species. They are herbivores, but the damage they do to vegetation is rarely noticeable. Horseback riders do, however, have unkind words to say about the burrows they make.

Richardson's name is also attached to subspecies of the boreal owl, the blue grouse, and the merlin (or pigeon hawk), as well as plants such as *Geranium richardsonii* and others. Although Richardson suffered physical hardships on his two trips with Franklin, they could scarcely compare with the emotional stress he must have experienced when, at the age of sixty-two, he embarked on a futile effort to discover what had become of Franklin's third expedition to the Arctic.

Ross's Gull

Richardson also wrote up the results of the arctic explorations of James Clark Ross (1800–1862). Ross is remembered by bird lovers as the discoverer of the elegant and distinctive Ross's gull, the first specimens of which he shot on the Melville Peninsula, at the north end of Hudson's Bay, in 1823. These gulls, in breeding plumage, have a narrow black ring around their neck and over the back of their crown, rendering them much easier to identify than most gulls. Unfortunately you or I may never see one, since these gulls spend their entire lives in the Far North. No one knew where they nested until 1905, when a Russian explorer found a breeding colony in Siberia. Since then a few nests have been found in northern Canada. It was Richardson who named the gull *Larus rossii*. Unfortunately he delayed publication, and in the meantime William MacGillivray had obtained a specimen and named it *Larus roseus*. Priority demands that we use MacGillivray's scientific name, but the common name Ross's gull is still in use.

James Clark Ross went to sea when he was only twelve as a companion to his uncle, John Ross. In 1818 the two Rosses set out to look for the Northwest Passage under orders from John Barrow, secretary of the admiralty. Edward Sabine was with them; a small, black-headed gull they took came to be called Sabine's gull. This expedition returned when they failed to find a passage among the islands. James Ross next accompanied Sabine on an expedition led by Captain William Parry. This time the ships got much farther, spending a winter in Melville Sound, many miles due north of the coastline that Franklin was exploring that same year. In 1823 James Ross was on a second expedition headed by Parry, and it was on this trip that Ross's gull was taken.

James Ross's uncle John, in the meantime, was planning a privately financed search for the Northwest Passage. On the *Victory*, the first steamship to be used in arctic exploration, the two Rosses and their crew reached the Boothia Peninsula, west of Baffin Island, where they spent the winter. James used his rifle to supply the men with fresh musk ox, bear, and ptarmigan. The following summer they remained locked in ice, and James left by dogsled to locate the North Magnetic Pole. After a second and then a third winter locked in ice, they abandoned their ship in the hope of reaching Baffin Bay, five hundred miles away. After several died from scurvy before they reached the bay in August 1833, the survivors struck out into the bay in small boats. Fortunately they were soon rescued by a whaler. Back in England, they had long since been given up as lost. The more or less happy ending of their long period in the deep freeze stands in contrast to the fate of Franklin's expedition a decade later.

James Clark Ross was now only thirty-seven years old, and his life of adventure was far from over. He had survived eight winters in the Arctic and had found the North Magnetic Pole. Thus he was a logical choice to locate the South Magnetic Pole and explore the almost unknown Antarctic continent. He set out with two ships, the *Erebus* and the *Terror* (the same two ships that would later carry Franklin and his men to their death in the Arctic). On board as naturalist was Joseph Dalton Hooker (1817–1911), son of W. J. Hooker and like him associated with the Royal Botanical Gardens at Kew. Hooker had just received a copy of Darwin's *Journal* from his voyage on the *Beagle* from its author, and he carried it with him. After visiting John Franklin in Tasmania, the ships headed south, discovering what is now called the Ross Sea, Ross Ice Shelf, and Ross Island, at the same time naming a series of peaks on the mainland: Mount Barrow, Mount Sabine, Mount Erebus, and the Franklin Mountain Range. Eventually they circumnavigated the continent, Ross recording in his journal the habits of penguins, petrels, and other antarctic birds. He made space in his cabin for Hooker to study and draw plants, and on his return Hooker published *Flora Antarctica* (1847), the first review of the plants of Antarctica, New Zealand, and Tasmania.

Back in England, Ross was asked to undertake another search for the Northwest Passage, but he declined and Franklin went instead. But when Franklin failed to return he, like Richardson, joined in the search, without success. It was his last expedition.

He had hoped to spend his remaining years working on his notes and collections, but his wife's death left him with four children to bring up, and he never found time for science.

Ross's Goose

It might be expected that another arctic bird, Ross's goose, *Chen rossii*, was named for James Clark Ross, but that is not the case. John Cassin named the goose for Bernard Rogan Ross (1827–1874), an officer in the Hudson's Bay Company who cooperated with Robert Kennicott in collecting specimens to send to Spencer Baird at the Smithsonian Institution. Ross's goose is sometimes called the lesser snow goose, and it sometimes hybridizes with members of that species.

Rudbeckia

It is a bit confusing to have three Rosses to consider, and nearly as confusing to have two genera of showy, sunflowerlike plants named *Rudbeckia* and *Rydbergia*. Olof Rudbeck (1660–1740) was a professor at the University of Uppsala in Sweden during Linnaeus's years as a student there. Rudbeck recognized Linnaeus's genius and appointed him to lecture to his botany classes and to tutor three of his twenty-four children. Linnaeus was later appointed to Rudbeck's professorship. After describing *Rudbeckia*, Linnaeus wrote to his old professor: "So long as the earth shall survive, and each spring shall see it covered with flowers, the *Rudbeckia* will preserve your glorious name." *Rudbeckia* is a genus of tall plants with large yellow flowers. The common names tall coneflower and black-eyed Susan are often applied; cultivated varieties are called golden-glow.

Rydbergia

Per Axel Rydberg (1860–1931) was also a native of Sweden, but he came to America in 1882 and became a field agent for the U.S. Department of Agriculture, later for the New York Botanical Garden. This gave him much opportunity to travel in the West, and in 1900 he wrote a *Catalogue of the Flora of Montana and the Yellowstone National Park*, followed by *Flora of Colorado* (1906) and *Flora of the Rocky Mountains and Adjacent Plains* (1917). Altogether he described at least a hundred genera and seventeen hundred species, but he was a something of a "splitter," with a tendency to call local variants "species" and to use genera of very limited scope. He and Aven Nelson of the University of Wyoming constantly argued over the status of various genera and species. But they were friends, and Nelson named a beard-tongue for him, *Penstemon rydbergii*. Rydberg and Nelson both provided important introductions to the exciting flora of the Rocky Mountain area.

It was Edward Lee Greene (1843–1915) who described *Rydbergia*. Greene went to Colorado in 1870 to collect plants for Asa Gray, but he later acquired enough self-confidence to maintain that botanists should publish on their own and not kowtow to Professor Gray of Harvard. Greene had unconventional ideas on the classification of plants, and he was an extreme splitter like Rydberg. Gray was only one of several botanists who condemned his research. Greene became a professor at the University of California in 1885 and remained there for ten years. He was an imposing figure on the lecture platform, with handsome features, a shock of white hair, and an eloquence that reflected his part-time employment as a preacher. Late in life he moved to Catholic University in Washington and became an associate of the Smithsonian Institution. Gray named the genus *Greenella* for him; it is a white-flowered, daisylike plant of the Southwest. Greene's blue-eyed Mary is also named for him.

From my desk, I can look into the valley below, where *Rudbeckia* stands tall and golden in midsummer, and off to the tundra in the distance, where I know *Rydbergia* will be blooming, its ragged yellow blossoms earning it the popular name "old man of the mountains." I have finally learned not to confuse the two names.

Say's Phoebe

$\mathcal{S}$pring comes reluctantly to the High Plains, the Great
Basin, and the foothills of the Rockies. There are bright,
warm days and a few bold spring flowers; then a cold front
passes through, with snow and biting winds. Even so, in March,
the first Say's phoebes (*Sayornis saya*) will be back, sitting on
bushtops or fence posts, twitching their tails and flying off to
snag whatever insects may be on the wing so early in the season.
These are males, looking for nesting sites or, more often than
not, returning to the place where they nested the year before.
Now and then one will be caught in a snowstorm. An observer
watched one in a late February storm and pitied its "fluffed
loneliness as it sat on a limb heavily laden with snow." This individual was seen to snatch berries from an ivy that apparently
sufficed until insects were flying again.

In a week or so the females will be back, and pairs will settle
in a sheltered place to build their nest. Probably they originally
adopted protected rocky ledges or cavities in trees, but nowadays they often build under bridges or the eaves of buildings.
Like eastern phoebes, they seem well adjusted to living near
people, and many a western ranch has its resident pair. These
are handsome birds, slightly larger than eastern phoebes and

191

Say's Phoebe

tawny rather than white beneath. Their song is a sharp *pit-tsee-ar*, while their call is a plaintive *phee-ear*, rather melancholy, as perhaps befits the aridity of the country they inhabit.

Nests are built of grasses and plant fibers and lined with fine materials such as cow hairs. Four or five white eggs are laid and incubated by the female while the male sits close by. When the eggs hatch, both parents bring in beakfuls of insects every few minutes. Usually there is a second brood each summer. There are several records of Say's phoebes taking over the nests of other birds, especially barn swallows and cliff swallows.

As is typical of a member of the flycatcher family, Say's phoebes live almost exclusively on insects they catch on the wing. Foster Beal, who worked under C. Hart Merriam at the U.S. Biological Survey, in the course of his career examined the stomach contents of untold thousands of birds in the effort to evaluate their importance to farmers. He found that bees and wasps made up the highest percentage of the prey of Say's phoebes (31 percent), followed by flies (17 percent), and grasshoppers and crickets (15 percent), with smaller percentages of a wide variety of other insects.

Flycatching behavior occurs among members of many groups of birds, not all related to phoebes (Australian flycatchers belong to the thrush family). Phoebes belong to a family of strictly American distribution called the tyrant flycatchers. It was Linnaeus who first applied the word *tyrant* (Latin *tyrannus*) to these birds, possibly because they tend to perch on high objects as if dominating the landscape. Tyrant flycatchers make up the largest family of American birds. There are nearly four hundred species throughout the Americas, the smallest no more than two and a half inches long, the largest more than a foot. It is a diverse group, including the kingbirds, the kiskadees, and the graceful scissor-tail, as well as many smaller species garbed mostly in browns and grays. Some of the many tropical species snatch berries and fruits from trees as well as flying insects. The great kiskadee sometimes takes snakes and frogs and has even been seen to dive into shallow water to take tadpoles and small fishes. This is a Central American species that sometimes ranges into extreme southern Texas.

Male and female Say's phoebes are virtually identical in coloration, and this is true of nearly all flycatchers, though in striking contrast to many birds in which males are much more brightly colored than the females. The ornate plumage of many male birds is the product of sexual selection; females choose superior males on the basis of their bright feathers and the vigor of their displays. But, as noted in the case of Gambel's quail, these colors and displays render the males conspicuous to predators, resulting in counterselection depending upon the kind and abundance of predators. Most flycatchers are monogamous, and males tend to remain close to the nest, feeding the young and later teaching them to catch insects. Under these conditions, loss of the male would be a threat to the family, and natural selection evidently had deterred any tendencies to develop bright colors and elaborate displays.

It is always risky to generalize about nature, and one species of flycatcher provides a notable exception to what I have just said. This is the vermilion flycatcher of our Southwest. The females are rather ordinary-looking small birds, but the males have a vivid scarlet cap, throat, and breast. I remember stopping on a day in May along the Gila River in New Mexico, where the trees were filled with these sparks of fire, apparently just returned from their wintering sites in Mexico. The species name is particularly fitting: *Pyrocephalus rubinus* (firehead of reddish color). Arthur Cleveland Bent described the courtship of the male as follows:

> Starting from his perch on the top of some tall weed stalk, or low dead branch, he mounts upward 20, 30, or even 50 feet, in an ecstasy of excitement, the fiery crest erected, his glowing breast expanded, his tail lifted and spread, and his wings vibrating rapidly, as he hovers like a sparrow hawk in rising circles; at frequent intervals he pours forth a delightful, soft, twittering, tinkling love song, all for the delectation of his chosen mate, clad in somber colors and hidden in the foliage below; then slowly he flutters down to claim her. . . .

How does it happen that vermilion flycatchers depart so dramatically from the way of life of most flycatchers? The males are very aggressive in defense of their territories, and it may be that they compensate for their increased visibility to predators by their feistiness and their flashy colors—for red is a warning color understood by many birds. William Beebe apparently felt that this might be the case. "This beautiful creature [he wrote] must have had some talisman which guarded him from the fate which overhangs brilliantly coloured birds, for he seemed to have no fear of showing his beauty. . . . Although we watched long and carefully, we never saw a Vermilion Flycatcher assailed or threatened by a shrike or hawk."

The much more modestly colored phoebes (genus *Sayornis*) include not only Say's but also the black phoebe of the Southwest and the familiar eastern phoebe (which is the only one that actually sings *fee-bee* and thereby gives the genus its common name). Thomas Say (1787–1834) discovered the species that bears his name in 1820, when he was exploring southern Colorado as a member of Stephen Long's expedition to the West. Say is most often thought of as an entomologist, and it is true that insects were his first love, but he also collected and described birds, mammals, snakes, and shells, and he made contributions to the study of the customs and languages of Native Americans. This was a considerable accomplishment for a person of limited education who never had much money and was ill off and on for much of his life.

Say had the good fortune to be born in Philadelphia, then the intellectual capital of the country, and to be the son of a physician and state senator, the great-grandson of naturalist John Bartram and grandnephew of William Bartram. In Bartram's gardens he met Alexander Wilson and other naturalists, and at Peale's Museum he met Titian Peale and his brothers as well as Thomas Nuttall. In 1812 he became a member of the Philadelphia Academy of Natural Sciences, founded that same year. When a brief career as a druggist failed, Say came to spend much of his time at the Academy and the Museum. Say and Nuttall often worked late into the night, sometimes sleeping in the only place in Peale's Museum large enough to accommodate them: beneath the mastodon skeleton.

In 1817 William Maclure joined the Academy, "a gentleman [in Say's words] well known in Europe & America for Science and beneficence." Maclure remained his patron for much of Say's life. In the fall of that same year Maclure, Say, Peale, and George Ord set off to Georgia and Florida, but they returned before spring with limited collections because of the threat of Native American attack. A much greater opportunity greeted Say in 1819, when he was appointed zoologist of Major Stephen Long's western expedition. Titian Peale was also a member of the party, as assistant naturalist. Say was then thirty, Peale eighteen.

The first summer was spent traveling down the Ohio and up the Missouri in a steamboat. It was relatively unproductive from Say's point of view, but during the winter encampment, near Council Bluffs, Iowa, he made significant collections of birds, mammals, and insects. The following summer, joined by botanist Edwin James, the expedition headed for the Rockies, meeting the mountains near the present site of Denver, following them south to the site of Pueblo, then returning along the Arkansas River. Long has often been criticized for failing to find the source of the Red River, as he had been directed to do, and for characterizing the western plains as the "Great American Desert." Many of the specimens collected arrived back in poor condition, and many of Say's notes were lost when taken and apparently destroyed by deserters. Nevertheless the scientific accomplishments of the expedition were considerable. Long mapped previously unknown areas, James collected many new plants (most of them described by John Torrey), and Say collected and later described thirteen new mammals, as many birds, several reptiles and amphibians, and several hundred insects.

Say soon accompanied Long on a second expedition, this one to Minnesota and adjacent parts of Canada, returning via the Great Lakes. Despite the loss of some of their collections, Say was able to describe several fishes, leeches, shells, and insects. Back in Philadelphia, he corresponded with many naturalists of the day and also helped Charles Bonaparte complete his supplement to Wilson's *American Ornithology.*

In 1824 Say published the first part of his *American Entomology,* which was designed to parallel the *Ornithology* of his friend Wilson. "The author's design [he explained in his preface] is to exemplify the genera and species of the insects of the United States by means of colored engravings." Over time fifty-four colored plates appeared, many of them done by Titian Peale. Unfortunately only three of the many proposed parts of *American Entomology* appeared, but Say published in scientific journals for the remainder of his life. John L. LeConte compiled all of Say's publications on insects and published them in two volumes in 1859.

Say was not to stay in the stimulating atmosphere of Philadelphia for long. In 1825 Maclure persuaded him to join him in Robert Owen's utopian project at New Harmony, Indiana. Say continued his research there and began an *American Conchology.* However, New Harmony's harmony did not last, and more and more he found himself involved with responsibilities he did not really enjoy. In 1826 he eloped with his artist, Lucy Sistare, "the handsomest and most polished of the female world" in New Harmony. The following year he was off to Mexico for several months with Maclure. But Say's health continued to be fragile, and in 1834 he succumbed to a fever, his work far from complete. Lucy Say lived another fifty-two years.

In an address to the American Philosophical Society shortly after Say's death, George Ord spoke of Say's "extraordinary exertions" on behalf of science and his readiness to "attend to the wants of others."

"His disposition [Ord went on] was so truly amiable, his manners were so bland and conciliating, that no one, after having once formed his acquaintance, could cease to esteem him. A remarkable feature of his character was his modesty, which, leading to habits of retirement, in some respects unfitted him for the intercourse of society, except that of his private friends, where, it may be said, he was truly at home, and where he was the idol of every heart."

The ever-critical Ord also pointed out Say's somewhat limited education and what he considered his shortcomings as a writer and his lack of "technical precision." Lucy Say resented these remarks and staunchly defended her husband in a letter to Massachusetts entomologist T. W. Harris. In her recent biography, Patricia Tyson Stroud has convincingly portrayed Say not as "bland" and "conciliating," but as "a strong, self-determined, highly motivated, even driving character." In time he taught himself all he needed to know as a naturalist, and considerably more, and it is amazing what he accomplished during a life of uncertain health mostly spent far from centers of learning. He was sometimes critical of other naturalists, including Rafinesque, who described new species at a furious pace and sometimes described the same ones as Say—often a bit earlier. He was also critical of those who sent their specimens to Europe for identification, for he believed strongly that the time had come for American self-sufficiency among naturalists.

Since Say was the first zoologist to penetrate the West as far as the Rockies, it is not surprising that it was he who described some of the West's most familiar birds, mammals, and reptiles. They include the coyote, the kit fox, the Colorado chipmunk, the band-tailed pigeon, the blue grouse, the pilot black snake, and others. Perhaps the best known of the hundreds of insects he described was the Colorado potato beetle, which became a scourge not only in America but in Europe as well. Besides the phoebes, Say had many insects, mollusks, and a variety of other animals named for him. One of the insects, a sand wasp, *Bembix sayi*, has played an important role in my own research.

Since Say's phoebe and the genus *Sayornis* were described by Charles Bonaparte, who will be the major figure in a later chapter, I'll take this occasion to say a few words about Titian Ramsay Peale (1799–1885), Say's friend and his companion on the Florida trip of 1818 and the Long Expedition of 1819–1820. Titian's father, Charles Willson Peale, was an artist and founder of Peale's Museum in Philadelphia. His sons, Titian, Raphaelle, Rembrandt, and Rubens, were all trained in art, but only Titian attained fame as a naturalist. Following the return of the Long Expedition, Titian was appointed assistant manager of his father's museum. He wanted to keep the museum open on Sundays, but the press was violently opposed to this misuse of the Sabbath. So he posted a sign in front of the door each Sunday morning:

> Here the wonderful works of the Divinity may be contemplated with pleasure and advantage.
> Let no one enter today with any other view.

In 1832 Peale traveled to northern South America to collect birds for Charles Bonaparte. Then, in 1838, he was appointed naturalist to the U.S. Exploring Expedition, headed by Lieutenant Charles Wilkes. With a fleet of six ships, this expedition spent nearly four years at sea, mostly in the Pacific, surveying 280 islands and part of the coast of Antarctica. Other notable naturalists on the expedition were Philadelphia physician and botanist Charles Pickering and Yale geologist James Dwight Dana. Besides collecting over two thousand birds and over a hundred mammals, Peale made many excellent sketches and paintings, supplementing those of the expedition's two official artists. This was, of course, before photography was well advanced, and illustrations made on the spot formed an important part of the records of expeditions of the time.

Peale was thirty-nine years old when the expedition started and had had much experience as a collector and preparator of specimens, but this was the first time he had played the role of a senior naturalist. On his return he prepared the expedition's report on birds and mammals, a report that included much valuable information—for example, on species of Hawaiian honeycreepers that are long since extinct. But his shortcomings as a systematist were apparent, and this volume was later withdrawn from publication. It was revised ten years later by Philadelphia ornithologist John Cassin, under his own name.

After his return from the Wilkes Expedition, Peale received an appointment at the U.S. Patent Office. The collections from the expedition were held and displayed at the Patent Office until they were removed to the Smithsonian in 1858. Peale had hoped for an appointment to the Smithsonian, but the position went to Spencer Baird. During the later part of his life, Peale moved back to Philadelphia and turned to photography, though he continued to paint. Peale's Museum had by this time failed, its contents having been sold at auction in 1849. Many of Titian Peale's paintings and photographs are still in existence. He described quite a number of birds from the Wilkes Expedition as well as several marine mammals, including Peale's dolphin of the North Pacific. Subspecies of the dusky dolphin and the peregrine falcon are named for him, a person who was always more at home in the field with a gun, a notebook, and a paintbrush than in the halls of science.

Steller's Jay

When Thomas Say traveled down the Front Range of the Rockies as a member of Long's Expedition, he kept mostly to the plains and foothills, leaving it to the younger Edwin James to climb Pike's Peak and to explore the Arkansas Valley as far as

Royal Gorge. So he may never have seen one of the Rockies' most striking birds, Steller's jay, which prefers to live in forests of ponderosa pine at moderate elevations. These birds have the longest crests of any North American birds, composed of feathers that can be extended or depressed to suit their moods. Their plumage is a demonstration of shades of blue: breast, wings, and tail shades of indigo, grading into a deep blue-black on the head and upper back. White streaks about their eyes give the birds a roguish appearance that suits them well. Unfortunately they are not musicians, but they do at least produce an interesting variety of sounds, some of them resembling the calls of red-tailed hawks. In summer they eat a great many insects and steal a few birds' eggs, and in winter they supplement a diet of pine seeds with whatever they can filch from bird feeders. So adaptable are they that they thrive all the way from Alaska to Guatemala, always in forested mountain country. Elliott Coues described them well: "a tough, wiry, independent creature, with enough sense to take precious good care of himself."

The jays are named for Georg Wilhelm Steller (1709–1746), who discovered them in 1741. Steller was a German naturalist who was attached to a Russian expedition that left Kamchatka in two ships to explore the "Great Land" to the east, now called Alaska. It was headed by Vitus Bering (a Dane), for whom the Bering Sea is named. Steller discovered not only the jays but several other creatures that were later named for him: Steller's eider duck, Steller's sea-eagle, and Steller's sea cow (now extinct). He found the jays during a landfall on Kayak Island (not far from present-day Valdez). He recognized them at once as relatives of the blue jays of eastern North America, which had been described a few years earlier. "This bird [he said] proved to me that we were really in America."

In fact Steller was able to spend only one day on the Alaska coast before Bering set sail for home. But they were busy hours and resulted in the first scientific report on the plants and animals of Alaska. In the words of a biographer, "Perhaps no other naturalist in history ever accomplished so monumental a task under such difficulties and in so little time." The return trip proved disastrous. Violent storms constantly drove the ships off course, and it was late November before they approached the Asiatic continent. In fact they did not make it, for they were shipwrecked on what is now called Bering Island. Bering died there, and most of the officers and crew were ill with scurvy. Remarkably, Steller and several others survived the winter and built a new ship from the remains of the old. They arrived safely in Kamchatka the following summer. But Steller was plagued with bad luck and died a few years later while wandering around Siberia, still finding plants and animals new to science. Fortunately his journals survived, so we may still share in his adventures.

Scott's Oriole

Scott's oriole is another notable western bird, much more addicted to arid regions than either Say's phoebe or Steller's jay. Males display a brilliant yellow and black plumage and advertise their presence with a song of rich, liquid phrases. This is perhaps the only bird named by an army general for another army general. General Winfield Scott (1786–1866) spent his entire career in the army and emerged from the Mexican War a national hero. He knew every president from Jefferson to Lincoln and was himself once nominated for the presidency. To his troops and fellow officers, he was "old fuss and feathers," so punctilious was he in dress and decorum.

There is no evidence that Scott was an avid bird-watcher or found much time for natural history. Not so Lieutenant (later General) Darius Nash Couch (1822–1897), who served under

Scott in the Mexican War and, despite a lingering illness contracted during that war, went on to distinguish himself in the
Civil War. Couch took a leave of absence in 1853–1854 and
made an expedition into Mexico to collect specimens for the
Smithsonian Institution. Spencer Baird wrote to his father in
1853: "Lt. Couch U.S.A. left in January for a trip in Northern
Mexico. He expects to stay a year and to make huge collections
of all sorts of critters. He has already sent in to us some valuable
things. . . ."

Among the "critters" he collected was an oriole he named
Icterus scottii. In his dedication, Couch wrote: "I have named
this handsome bird as a slight token of my high regard for Major
General Winfield Scott, Commander in Chief of the U.S.
Army." The oriole was found to have had an earlier scientific
name, but the common name Scott's oriole is still used.
Couch's kingbird, *Tyrannus couchii*, and Couch's spadefoot
toad, *Scaphiopus couchii*, were both described from his Mexican collections by Baird. Subspecies of the rock squirrel and the
northern pocket gopher also bear Couch's name.

Townsend's Mole

*M*oles have some of the finest fur in the animal world, and at one time moleskin was popular for men's and ladies' coats and hats. But moles have never had good press in spite of the endearing qualities of Mole in Kenneth Grahame's book *The Wind in the Willows*. Few people have ever seen one, though many have seen their unwelcome channels beneath lawns and golf courses. They do not make good pets as they disdain light, and when two moles are placed in the same cage they may fight to death. For the same reasons they are difficult to study and largely remain creatures of mystery.

Kenneth Mellanby, in his book *The Mole*, tells us that King William III of England was killed in 1702 when his horse stumbled over a molehill. While this may have been an isolated incident, moles have long been condemned because of their habit of disfiguring sod and disturbing the roots of garden plants. They do not, however, feed on plant roots but on earthworms, wireworms, and other small soil animals. In the final reckoning, the benefits moles provide by turning over soil and destroying destructive soil insects may at least equal the damage they do.

Moles require relatively moist and loamy soil and are there-fore absent over vast areas of the earth, including many parts of the western United States. There are only twenty species of moles in the world; three are in the eastern United States and four others along the West Coast. Townsend's mole (*Scapanus townsendi*) is the largest in North America, six to seven inches long, with a tail adding another two inches. John Kirk Townsend (1809–1851) found it at Fort Vancouver, Washington, while he was acting surgeon there in 1835. It is restricted to moist soils of the far Northwest; it hasn't been well studied, but presumably it doesn't differ very much from other, better-studied species.

Moles are nearly blind, and their tiny eyes probably do not see images but can tell light from dark. There are no external ears, which might prove a handicap in the confines of the burrow. But internal ears enable them to hear pretty well, and captive moles have been taught to come for feeding at the sound of a bell. Their snout is tender and more or less hairless, but there are small papillae and long bristles, which are very sensitive to touch and help guide the moles through their underground mazes and to food. Probably they are sensitive to subtle odors in the soil, too. Moles require about half their own weight in food each day. They dig and feed in bursts of activity, with periods of rest in between.

Townsend's Mole

This was shown by attaching capsules of radioactive cobalt to moles and following them beneath the soil with a Geiger counter.

In summer, moles burrow just beneath the surface; in cold weather they retreat to deeper burrows, where it is warmer and there is more food. Like many mammals, moles have a home territory, but in spring males may make long burrows in the search for a female. Young are born in an underground nest lined with grasses and leaves. The young grow rapidly and within a month are nearly full size. They soon depart, often above ground, to find territories of their own.

John Kirk Townsend was a Philadelphia-born Quaker. Several members of his family were bird-watchers, and John learned the art of taxidermy at an early age. The Townsends knew Audubon well, and when Audubon stopped in Philadelphia in 1833 he found that John's "zeal for the study of ornithology was unrelented." Townsend showed Audubon a bird he had recently collected near Philadelphia and a description he had drawn up for it. He proposed calling it Audubon's bunting, but in fact it was Audubon who formally described it, reversing the compliment and calling it Townsend's bunting (*Spiza townsendii*). This bird has never been seen again. It did not appear to be a freak or a hybrid and has remained a puzzle to ornithologists ever since. Perhaps Townsend had collected one of the last specimens of a species approaching extinction.

When he was twenty-five, Townsend had an opportunity to join Nathaniel Wyeth's expedition to the Pacific, with botanist Thomas Nuttall as a companion. Townsend's account of the trip, *Across the Rockies to the Columbia*, is one of the most enjoyable stories of western exploration and has recently been reissued in paperback. The harvest of new birds and mammals (and for Nuttall, of plants) was great indeed, but there were times when specimens were lost to science. One morning Townsend shot an interesting owl which he intended to stuff in the evening. But on returning to camp in the afternoon he found that Nuttall and a companion had roasted the owl and were "picking the last bones." Townsend carried a two-gallon bottle of whisky to preserve specimens, and ascending the Willamette River, he left the bottle on board the skiff while he explored the countryside. When he returned he found that the expedition's tailor had "decanted the liquor from the precious reptiles which I had destined for immortality, and he and one of his pot companions had been 'happy' upon it for a whole day."

After reaching the Pacific, Townsend went for a time to the Hawaiian Islands, then returned to Fort Vancouver, where he acted as post surgeon for several months, since the regular surgeon had left. He returned east via Cape Horn. In 1842 he became associated with the National Institute, something of a forerunner of the National Museum, which was housed in the Patent Office. Spencer Baird visited Townsend and his wife, Charlotte, and regarded them as "the cleverest people of my acquaintance" (Charlotte's sister had married Baird's brother). As a curator at the National Institute, Townsend often bought specimens in local markets and mounted them, along with those from various exploratory expeditions. He was a superb taxidermist and had a special powder of his own formulation that he used to prevent damage to specimens from museum pests. The powder contained arsenic, and it was not long before Townsend's health began to decline. He died of chronic arsenic poisoning in 1851, only forty-two years of age.

Townsend's collections from his western trip proved a bonanza to naturalists. The Easter daisy of the West was named *Townsendia*; Townsend's warbler is an attractive bird of coniferous forests, Townsend's solitaire a modestly colored but superb vocalist widely distributed in the West. Townsend called the solitaire a "new species of mocking bird," but it is in fact a thrush. Among the mammals he collected was Townsend's big-eared bat, Townsend's ground squirrel, the white-tailed jackrabbit (*Lepus townsendii*), and, of course, the mole.

Description and naming of Townsend's mammals fell to John Bachman (1790–1874), who was then cooperating with Audubon on the three-volume *Viviparous Quadripeds of North America*. Bachman had, in his own words, "from my earliest childhood . . . an irrepressible desire for the study of Natural History." As a youth he trapped beavers around his home in New York state and sold the skins to buy books on birds and mammals. For a time he taught school in Philadelphia, where he met Alexander Wilson.

Bachman came from a Lutheran family, and in 1814 he was ordained a Lutheran minister. He was sent to Charleston, South Carolina, where over time he built up a strong congregation. His sermons were evidently quite liberal for the time and were sprinkled with allusions to nature. He spent his spare time collecting birds and mammals from the Southeast and studying their life histories. Audubon visited him in 1831, the beginning of an association that lasted until Audubon's death, and beyond through Audubon's two sons, who had married Bachman's daughters.

Bachman struggled to reconcile his religious beliefs with concepts of science. During the Civil War, he was similarly torn between two extremes, as he was a Northerner by birth yet had many friends in the South, including many African Americans. Eventually he lost much of his property because of his equivocation. Audubon named Bachman's sparrow, Bachman's warbler, and the black oystercatcher (*Haematopus bachmani*) for him. The brush rabbit (*Sylvilagus bachmani*) also bears his name. Bachman was widely admired for his integrity and vision, and his name is perpetuated in those of several notable birds and mammals.

Traill's Flycatcher

The relationships of the volatile Audubon with his contemporaries were often complex. In a letter to his close friend Bachman he called Townsend "lazy and careless," since the field notes associated with his specimens were often sketchy and unreliable. Bachman was sometimes impatient with Audubon, complaining that he spent his days on the upper Missouri hunting buffalo when he should have been collecting mammals for their joint venture on quadrupeds. One friendship that was never strained was that of Audubon and Thomas Stewart Traill (1781–1862), a Scottish physician who had a practice in Liverpool at the time of Audubon's visit in 1826. Traill was much impressed with Audubon's paintings and helped arrange a highly successful exhibition at the Royal Institute in Liverpool and later a similar exhibition in Manchester. After Audubon went to London, Traill introduced him to Henry Bohn, a book dealer who gave Audubon valuable advice on preparing his paintings for publication. When Audubon went to Edinburgh, he carried letters of introduction from Traill to important people there. These were worrisome days for Audubon; he had left his family in America and was, as usual, short of funds. But with the help of Traill and others he accumulated subscribers to his *Birds of America* and obtained the services of Robert Havell.

Traill later moved to Edinburgh University as a professor of medicine. He was an editor of *Encyclopedia Britannica*, to which he contributed many articles. The bird Audubon named for him, Traill's flycatcher, was found to be a composite of two closely related species, one of which is now called the willow flycatcher, the other the alder flycatcher. The scientific name *Empidonax traillii* now applies to the willow flycatcher, a widely distributed but shy bird of orchards and thickets.

Tuckerman's Lichens

It is sometimes hard to think of lichens as living things, so much a part of the rocks bearing them they seem. Yet they are very much alive and play roles in nature that are not often appreciated. There are many kinds, each an intimate association between a species of alga and one of fungus. Together they form a complex pattern of black, brown, gray, green, or even orange, on rocks or trees or ground. The acids they produce erode rock surfaces, permitting higher plants to take root and water to enter crevices and freeze, all assisting in the breakdown of the rocks. Lichens serve as food for many animals. They have no deciduous parts, so they tend to accumulate poisons from the air and from rainfall; thus they are useful indicators of acid rain and aerial pollutants.

America's pioneer lichenologist was Edward Tuckerman (1817–1886), a member of an old New England family who spent much of his career as a professor at Amherst College. He spent many days roaming the wilder parts of New England collecting lichens as well as flowering plants (Tuckerman's Ravine in the White Mountains of New Hampshire is named for him). With encouragement from his friend Asa Gray, he published important monographs on lichens and was soon receiving lichens from various exploring parties in the West. The first volume of his *Synopsis of North American Lichens* was published in 1882; a second volume was completed after his death by one of his students. *Tuckermania*, a genus of composites, was named for him, as well as a sedge, *Carex tuckermanii*. He himself described several species of lichens, and surely appreciated that what some people consider the "trash" of the plant world—or ignore altogether—add a great deal of beauty to rocky landscapes and are far more important in the functioning of the natural world than they appear to be.

Uhler's Assassin Bug

*B*ugs do not touch the heart the way that furry mammals or feathery birds do, whether one means bugs in the vernacular sense of anything creepy-crawly, or in the entomological sense of insects having a particular kind of piercing-sucking beak. When John Southall wrote his *Treatise of Buggs* in 1730, it was bugs in the entomological sense he was thinking of, more specifically of bedbugs. Assassin bugs are kin to bedbugs but primarily out-of-doors bugs that prefer eating other insects to dining on people. There are about one hundred and fifty species of assassin bugs in North America. Uhler's assassin bug (*Barce uhleri*) is less than half an inch long, not big enough to assassinate anything much larger than a plant louse. It lives among grasses or sometimes in old barns or cellars, stalking small insects, grasping them in its front legs, and sucking out their blood with its stout beak.

There would perhaps be little to say about Uhler's assassin bug if it were not such a beautifully adapted creature. It is slender and sticklike, with extremely long legs hardly thicker than a thread; in fact, these insects are often called thread-legged bugs. The front legs are fitted for seizing prey; the basal segment (coxa)' is very long, and the following two segments (femur and tibia—names adopted from human anatomy) are spiny and fit together like a trap. These two features—sticklike body form and traplike front legs—have evolved independently in other groups of insects, for example in preying mantids and in water scorpions. Sticklike body form also occurs in walking sticks, some grasshoppers, and the larvae of certain moths (sometimes called measuring worms). Traplike front legs occur not only in preying mantids but in a group of quite unrelated insects that are spider parasites as larvae: a group called mantispids because of a superficial resemblance to mantids. Traplike front legs of somewhat different form also occur in a group of parasitic wasps, called dryinids, that sneak up on leafhoppers and grasp them in the forelegs while they lay an egg inside their prey.

These are all examples of convergent evolution, that is, instances of unrelated animals evolving, through natural selection, similar structures that are effective in meeting similar life exigencies. Examples of convergent evolution are widespread among animals. Some of the most striking are among the marsupials of Australia, where several species have come to resemble, at least superficially, the placental mammals of other parts of the world. Wombats resemble woodchucks, the now-extinct Tasmanian wolves resembled true wolves, and so on; there is

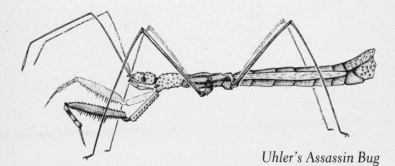

Uhler's Assassin Bug

even a marsupial mole, whose pouch conveniently opens backward rather than forward.

It is easy to see how useful traplike forelegs might be to a predator that stalks its prey. And after all, insects can easily walk on four legs, using the other two for some other function, as many do. Development of a sticklike form can also be advantageous to a predator, for the body casts little shadow and seems to be just another stick or grass blade. Plant feeders such as walking sticks and grasshoppers can also benefit from such a sticklike form, since birds and lizards may not recognize them as something edible. There is evidence that birds do in fact fail to detect stick-shaped caterpillars. Dr. L. De Ruiter of Oxford University found that jays could not discriminate between stick-shaped caterpillars and twigs. However, once a jay accidentally found such a caterpillar, it was quite able to catch others of the same kind.

Resemblance to sticks, grass blades, and twigs is by no means confined to insects. Potoos, tropical relatives of the whip-poor-will, rest in the daytime in an erect posture atop stumps and avoid detection by resembling an extension of the stump. Pipe fish, in tropical seas, drift about like bits of the seaweed in which they occur.

Biologists sometimes disagree as to whether these are examples of camouflage or of mimicry. In an interesting article entitled "A stick is a stick and not worth eating," Michael Robinson argues that sticklike animals are mimics (he calls them plant-part mimics), since they are providing false information to a predator rather than hiding by blending with their background. Robinson spent many years studying insects in Panama with the Smithsonian Institution's Tropical Research Institute before assuming the directorship of the National Zoological Park in Washington, D.C.

No one has studied Uhler's assassin bug to determine how effective its body form is in enabling it to sneak up on its prey or to avoid being eaten by birds or other predators, but one would guess that it goes about its business unnoticed except by keen-eyed entomologists. Philip Reese Uhler (1835–1913) was the first person to collect specimens and place them in a museum, Agassiz's Museum of Comparative Zoology at Harvard, where he was at the time curator and librarian. Uhler remained there three years (1864–1867), then departed abruptly for unexplained reasons. Agassiz had recently experienced a rebellion among his students because of strict regulations he had imposed, and several excellent young naturalists had left. Whether Uhler also resented Agassiz's authoritarian ways is unrecorded.

Uhler was born in Baltimore, Maryland, son of a well-to-do businessman. He began collecting butterflies and moths when he was only ten years old, and even after his father set him up in business he spent most of his time studying natural history. After his brief stint at Harvard, he became librarian of the Peabody Institute in Baltimore. Later he was involved in the founding of Johns Hopkins University, and he was a professor there until his death. His travels were somewhat limited, mainly in the eastern United States and the West Indies. However, he had an opportunity to study insects collected by the Hayden Survey and the Northwest Boundary Survey.

It was apparently during his studies of fresh material from the western states that he became specially interested in sucking bugs; perhaps he felt that more popular groups of insects, such as beetles and butterflies, were already in the good hands of people such as John L. LeConte and William Henry Edwards. His descriptions of the many new species he found were meticulously detailed, in contrast to the extremely brief descriptions of Thomas Say and other earlier entomologists. Uhler's life was "the quiet and uneventful one of the student; his profound modesty kept him in the background, and he disliked what he termed 'cheap notoriety.' . . . No worker appealed to him in vain, and to many he was of greatest help." (These are the words of L. O. Howard, a younger contemporary who was chief of the U.S. Bureau of Entomology for many years.)

As he grew older, Uhler suffered from diminished eyesight and had to give up his studies of insects. In the course of his career he described insects as diverse as cicadas, water bugs, and mole crickets. Uhler's assassin bug is only one of more than twenty-five species named for him, including a stink bug, which surely would not have bothered him a bit. Some of his collections ended up at the U.S. National Museum, others at the Museum of Comparative Zoology. It was a later curator at the MCZ, Nathan Banks (1868–1953), who discovered specimens Uhler had collected many years earlier in North Carolina and described Uhler's assassin bug.

Although Banks's life embraced the first half of the twentieth century, I do not hesitate to include him here, since he was devoted to the description of insect species and genera, and he had little interest in genetics, evolution, or other topics that excited his contemporaries. Also, by including Banks I can speak for myself, for he was one member of the "old school" of naturalists whom I knew personally, though only briefly. Banks published on a great many different kinds of insects, spiders, and mites; one of his favorite groups was the spider wasps. When I elected to study this group, as a graduate student at Cornell University, I had to study specimens at the MCZ, so I visited Banks. He seemed brusque and slightly formidable, but of course he gave me access to the collections. Doubtless he wondered if I were really a serious student, or if I might find flaws in his research. In any case, on my second visit he was much more cordial. He was already well past seventy, and he retired a year or two later. After retirement, he worked on a massive bibliography on the habits of insects, which remained incomplete when he died at eighty-four.

As one might predict, Banks had been interested in insects from childhood, and since he was born and brought up in New

York state, it was logical that he go to Cornell to study under pioneer entomology teacher John Henry Comstock. For over twenty years (1890–1916) he was employed by the Bureau of Entomology in Washington. He lived in Falls Church, Virginia, for much of that time. After his move to Harvard he lived on a ten-acre plot at Holliston, Massachusetts, twenty-five miles from Cambridge. Aside from brief trips to Panama and to the Smoky Mountains of Tennessee, he did most of his extensive collecting in his own backyard and adjacent fields and wood-lands. Both Falls Church and Holliston became well known to entomologists through the many species he found there. In 1945 I visited Falls Church, hoping to collect some of the many insects Banks had found there. I became lost in a maze of streets and houses and came back almost empty-handed.

Banks had no assistants to help handle the very large collec-tions at the MCZ, except for occasional help from one of his eight children. Most of his time was devoted to routine curato-rial duties, which he clearly enjoyed. Despite his gruff exterior, he had a good sense of humor. In the course of his long life he described many hundreds of insect species and quite a few gen-era; the number of species named for him must be well over a hundred, though I have not counted them. One of them, Banks grass mite, is listed as a major pest by the Entomological Soci-ety of America. So it may be his fate to be remembered for a tiny arachnid he described early in his career rather than for some of the giant and colorful tarantula hawk wasps that he described later on.

Underwood's Ferns

Naturalists who earn the adjective "great" are usually those who study birds, mammals, wildflowers, or other living things that attract much popular attention. But there are many smaller and less glamorous organisms, and many of them play roles in nature that are important, even essential, to other forms of life. Thus there is room for Palmer and Muhlenberg and their grasses, for Uhler and Banks and their insects—and for Lucien Marcus Underwood (1853–1907) and his ferns, liverworts, and fungi. Underwood was born in upstate New York and became a professor at Syracuse University, later at Columbia. Early in his career he published *Our Native Ferns and How to Study Them* (1881), later *Moulds, Mildews, and Mushrooms* (1899). Both went through several editions and did much to stimulate interest in those neglected plants. He attempted to put the classification of ferns on a sounder basis, with much success. He served on the board of directors of the New York Botanical Garden and played an important role in initiating the series *North American Flora.* His own contribution to the *Flora* was published after his death by suicide when he was fifty-four. Suicide is an unusual death for a naturalist; most are too engrossed in nature to indulge in self-analysis. Underwood is said to have been temporarily out of his mind as a result of overwork. His name is perpetuated in that of a club-moss, *Selaginella underwoodii,* and a sagebrush, *Artemisia underwoodii.* A fern he described himself is called Underwood's moonwort.

Virginia's Warbler

*S*pring is a sometime thing, and in the Rockies it is likely to be punctuated by cold fronts that bring icy winds and snow well into April, even into May or early June, depending on the altitude. Those of us who live in the foothills look forward to the arrival, in mid-May, of Virginia's warblers (*Vermivora virginiae*), flocking through to find their nesting sites on semiarid slopes among mountain mahoganies and serviceberries. These are not spectacular birds. The males, even in full spring plumage, are mostly gray, with yellow splashes on the upper breast and beneath the base of the tail. There is a white eye ring and a small reddish topknot, not easily seen in the field. Like all warblers, these are very active birds, so active that as soon as a person picks one up in his binoculars, the bird is somewhere else. In contrast to many warblers, these nest on the ground, where the female will lay three or four white eggs sprinkled with reddish brown. Both parents will feed the young with caterpillars, so easily found among the trees and shrubs in midsummer. In September they will migrate a relatively short distance, to semiarid mountains in Mexico not unlike those in which they bred in midsummer.

Virginia's Warbler

To bird lovers, warblers are a very special group, and it is a challenge to be able to identify the many species by sight or by their songs. Warblers, as we know them, are a strictly American group, more properly called wood warblers, to distinguish them from unrelated kinds in the Old World also called warblers. There are somewhat more than one hundred species, roughly half of which migrate into the United States and Canada during the summer. In terms of their migratory behavior, three groups of warblers can be recognized: those that migrate into the western states (like Virginia's, Townsend's, and about ten others); those that migrate into the eastern states (more than thirty species); and those that migrate into either the East or the West (Wilson's warbler, the yellow warbler, and half a dozen others). (The Great Plains are not the best warbler country, at least partly because of the scarcity of woodlands.)

Warblers are beautiful creatures, often flashing yellow, chestnut, or other bright colors as they flit about the tips of branches in the search for insects. Many of the species that remain in the tropics year-round are especially gaudy. The mangrove warbler of Central and South America is entirely yellow except for a reddish-chestnut head and throat; the red warbler of Mexico is wholly red except for a white cheek patch. The painted redstart of Mexico fortunately does migrate a short distance into our southwestern states, and I well remember being surprised by one, close up, while hiking through a New Mexico canyon; a sudden flash of black, white, and red that left me breathless.

Some warblers, such as Virginia's and the painted redstart, migrate a relatively short distance, while others migrate several thousand miles. The blackpoll warbler, for example, winters in northern South America and breeds in our northeastern states, Canada, and even Alaska. Then there are those that remain in the tropics, moving about only slightly from season to season. It is difficult to understand the variation in migratory behavior within one group of rather similar birds. Presumably the group had its origin in tropical or subtropical America, since none of the species spends more than three or four months away from a warmer climate.

All birds, in fact all animals and plants, have a tendency to move into unoccupied habitats. It is possible that as the North American glaciers receded and the whole of the United States and Canada became gradually warmer, certain species of warblers and other birds extended their ranges northward into areas where there was less competition for food and nesting sites. Birds require far more food in the breeding season, as there are hungry young to feed, and temperate regions provide an abundance of insects in the warmer months. Those species that were by nature more mobile, and perhaps had longer and more pointed wings, may slowly have come to occupy habitats in North America, moving back south as temperatures declined in the fall and food became scarce. Over many thousands of years, natural selection favored bird species in which migration was well established. Migrations are triggered by changes in day length and are so firmly established genetically that young birds who have never migrated before seem to know when and where to migrate, without guidance from their parents.

Of course it will never be possible to explain precisely why any species occupies a particular habitat either in summer or in winter; such secrets are locked in the past. But it need not deter us from enjoying the flush of warblers in our trees in the spring; it adds, indeed, to the wonder of it. Many of our warblers are declining in numbers year by year as habitats, both summer and winter, are destroyed. Future generations (if they care) may have to read about the wondrous flood of gay migrants that once made spring so special.

Army surgeon William Wallace Anderson (1824–1911) collected the first specimens of Virginia's warbler while he was stationed at Fort Burgwyn, New Mexico, in 1858. He had been sending material regularly to Spencer Baird at the Smithsonian. When he discovered a previously unknown warbler, he wished to have it named for his wife Virginia, and Baird obliged him by doing so.

Anderson was born in South Carolina; his father was a physician with an interest in natural history that he conveyed to his son. The younger Anderson graduated from the University of South Carolina and studied medicine at the University of Pennsylvania. He entered the army in 1849 and through the influence of Baird was assigned to the Pacific Railroad Survey. Like most army personnel, he moved about frequently, but Fort Burgwyn seems to have been his most productive assignment.

Anderson married Virginia Childs, the daughter of a brigadier general, in 1855. They soon set off for the West in an army wagon. Virginia took her piano, but it was not in the best of shape when they arrived in New Mexico. Fortunately her husband was good with his hands and managed to repair it. Despite the itinerant life of an army family, the Andersons had nine children. During the Civil War, Anderson served on the Confederate side; after the war he resumed sending specimens to the Smithsonian.

Some Other Army Naturalists

Nowadays it is hard to believe that an army officer would spend time watching birds or collecting specimens, but in the nineteenth century quite a number did just that. They were for the most part trained in medicine, as Anderson was, and most maintained contact with Spencer Baird, who often saw that they provided him with material from interesting places. In earlier chapters I discussed Darius Nash Couch, Elliott Coues, John Charles Frémont, Edgar Mearns, and John W. Gunnison; in later chapters Samuel Woodhouse, John Xántus, William Alexander Hammond, and Henry Crècy Yarrow will make appearances.

Others should be mentioned at least in passing. Charles Bendire (1836–1897), though born in Germany, served in the U.S. Army through the Civil War and beyond. After spending much time in the West, he wrote an important book, *Life Histories of American Birds* (1892). Bendire's thrasher is named for him. Adolphus Lewis Heermann (1827–1865), for whom Heermann's gull is named, was an officer with the Pacific Railroad Survey and added much to knowledge of birds of the Far West. John P. McCown (1815–1879) served in the army for nearly thirty years. He found a bird he could not recognize and sent it to George Lawrence, who described it and wrote that "it gives me pleasure to bestow upon this species the name of my friend, Capt. J. P. McCown, U.S.A." The bird was McCown's longspur, a handsome bird of the northern Great Plains that now appears to be declining in numbers.

Not all army men were attracted to birds, and not all were medical officers. Thomas Lincoln Casey (1857–1925) was a West Point graduate who was commissioned in the Engineering Corps. He devoted his life to the study of beetles, following the pioneering work of John L. LeConte and George Horn. Horn at first encouraged him but soon became disillusioned as Casey poured out vast numbers of "new species," based on minute differences in size, body shape, surface sculpture, and other features. He was one of the first to use a microscope (rather than a hand lens) to study beetles, and it is said that he described 9,400 species that he claimed to be new. In fact most coleopterists agree that he was an extreme "splitter" and that many of his species represent no more than normal variation in populations. Beetles were to him only a hobby, and he often did not bother to read the publications of others. When he retired from the army he settled in Washington, D.C., and upon his death he was buried with full military honors in Arlington National Cemetery. His beloved microscope was buried with him. The Casey Room at the U.S. National Museum houses his mammoth beetle collection. It is said that words of profanity have sometimes issued from the Casey Room when it is visited by modern-generation coleopterists.

Woodhouse's Toad

*T*oads are not endearing animals, covered as they are with "warts" and capable of producing, in the mating season, no more than a monotonous wail from the ponds where they breed. But there is much to be said for them as predators on insects and other small invertebrates. Doubtless Samuel Washington Woodhouse (1821–1904) was not in the least embarrassed by having a toad he collected in Arizona named *Bufo woodhousei* in his honor. We now know that Woodhouse's toad has a wide distribution; it is distinguished from other toads by the crests on its head and the pale stripe down its back. Sometimes these toads congregate beneath streetlights in towns, snatching up insects that are attracted to the lights. They grow gradually over several years, finally reaching a length of five inches or so if they manage to avoid being run over by a car or being swallowed by a snake. Like other toads, they are able to gulp air and inflate their bodies to the point that snakes can often not swallow them. But that is no help when they are squashed on a highway or when they wither away because their breeding sites have been destroyed.

In the spring the toads emerge from overwintering sites in the soil and migrate to shallow ponds, cattle tanks, creeks, or irrigation ditches. Here the males produce a loud "waaaah," augmenting the sound by inflating their throat into a balloonlike sac in the usual manner of toads and frogs. Males defend their calling sites by attacking intruders of the same sex. Males attempt to clasp and mount any moving object they encounter. If it happens to be another male, he is released shortly, but if it is a female he hangs on for many minutes, fertilizing the string of eggs she produces. It has been shown that in some species of toads larger males are most often successful in mating, since they can displace smaller males. Also, females tend to choose larger males. Since larger males are older and have therefore escaped being preyed upon for a considerable time, females may be ensuring that their offspring will have superior survival potential.

A size advantage has been shown in species that are "explosive breeders," that is, those in which all mating occurs within a very few days. But at least some populations of Woodhouse's toads breed over a long period of time, showing up in ponds in small numbers anytime in spring or summer. Brian Sullivan, working at Arizona State University, found this to be true in

Woodhouse's Toad

several ponds in that state. He marked 322 males for individual identification by toe-clipping, a simple operation that does not handicap the toad in any way. Each toad was weighed and his activities, call rate, and mating success recorded over several days. He found that in this case mating success was not correlated with size but with site tenacity and with the number of calls produced per minute. Females visited call sites and selected males with higher call rates. Perhaps they used call rate as an indication of male vigor and by natural selection chose these males to ensure greater vigor in their offspring.

Even though female Woodhouse's toads select mates with higher call rates, in dense aggregations the average call rate of males is depressed. That is, the more males there are, the slower they tend to call. When Brian Sullivan taped songs and played them back to males, he was able to cause males to reduce their call rate. When males are crowded, they also tend to call out of synchrony with nearby males, thus reducing the overlap of sound with that of their neighbors. Presumably these responses make it easier for females to evaluate and select individual males.

In recent years many species of toads, frogs, and salamanders have been declining in numbers. Amphibians require water for breeding, and wetlands are being lost throughout the country. It is also possible that many of the ponds in which they breed are becoming too acidic as a result of airborne acids from the burning of fossil fuels. The permeable skin of amphibians is particularly likely to absorb lethal particulates in the air. One wonders what the decline of amphibians may mean to the future of other species, including ourselves.

In Woodhouse's day there was little reason for concerns about the environment, and concepts of sexual selection were still some time in the future. It was enough to add another species to the known fauna. As a physician in Philadelphia, Woodhouse had been infected with the desire to study natural history through his contacts with Thomas Nuttall and other members of the

Academy of Natural Sciences. When Colonel J. J. Abert, Chief of
the Army Topographical Engineers, was looking for a doctor and
naturalist to accompany expeditions to the West, Woodhouse
was recommended to him. The first expedition he joined took
him to Indian Territory, still relatively unexplored for plants and
animals. Later he was a member of the Zuni River Expedition,
seeking to find a better route to California. Here he had many
opportunities to collect specimens in the Southwest, but his
eagerness once backfired. He encountered a particularly fine
specimen of a rattlesnake and held it down with his gun while he
picked it up behind the head, hoping to preserve it. But, he later
wrote, "I had too long a hold, when he threw round his head and
buried his fang in the side of the index finger of my left
hand. . . . The pain was intense." He sucked the wound, ligated
the finger, and applied ammonia; then, at the recommendation
of friends, he applied the "western remedy," consisting of half a
pint of whisky followed by a quart of brandy, taken internally. As
a dedicated scientist, Woodhouse kept a careful record of the
symptoms and the effects of various remedies that he tried.

Some months later, while warming himself by the campfire in
the morning, Woodhouse received a Yavapai Indian arrow in his
leg. Fortunately it was not a serious wound. In spite of these misfortunes, he brought back extensive field notes on various birds
and mammals as well as many specimens, including those of
Woodhouse's toad, Woodhouse's daisy (*Bahia woodhousei*), and
the tassel-eared squirrel, which he himself named for his superior,
calling it *Sciurus aberti*. Other species named by Woodhouse
include Ord's kangaroo rat, the desert pocket mouse, the white-
throated swift, the black-capped vireo, and Cassin's sparrow. The
last was named for his friend John Cassin, curator of birds at
the Philadelphia Academy of Natural Sciences for many years.

Woodhouse later was a surgeon in Nicaragua during explorations for an interocean canal. He died in 1904, one of the last
of a breed of surgeon-naturalists associated with army explo-

rations. The specimens he collected became the property of the Smithsonian Institution, the depository of all natural history objects owned by the government. It was Charles Girard (1822–1895), the Smithsonian's herpetologist, who named the toads *Bufo woodhousei*. It was a good way to perpetuate the memory of a man who devoted much of an adventurous life to natural history.

Girard had been one of Louis Agassiz's students at Neuchâtel, Switzerland, and when Agassiz moved to the United States in 1847 he brought Girard with him. Girard left Agassiz suddenly in 1850, accepting a position under Spencer Baird at the Smithsonian. Since Agassiz tended to look down on Baird, he resented Girard's move, and thereafter had little to do with him. Girard, though a bachelor of retiring nature, seemed to invite controversy. He was sympathetic to the South during the Civil War, and while visiting Paris he began sending medical supplies to the Confederate forces. This made it difficult for him to return to Washington, so he settled in Paris. During the war of 1870 he served as a French army physician and did important research on the causes of typhoid fever. Thus he made his mark on science in quite diverse ways.

Wislizenia *(Jackass Clover)*

Woodhouse had some harrowing experiences during his trips to the Southwest in the 1850s, but they were rivaled by those of Frederick Adolph Wislizenus (1810–1889), a German physician who fled to Switzerland and later to America following an unsuccessful student uprising in which he participated. He established a medical practice near St. Louis and became acquainted with George Engelmann. In 1839 Wislizenus "felt the need of mental and physical recreation" and joined a group of fur traders going up the Oregon Trail. He went as far as what is now Idaho, then returned with a small group through Colorado to Bent's Fort, on the Santa Fe Trail, thence back to Missouri. The book he wrote on his return, A *Journey to the Rocky Mountains in the Year 1839*, is still good reading, though as he admitted, he had neither the time nor the knowledge for much natural history.

A few years later, just before the outbreak of the Mexican War, Wislizenus joined a group of traders going to Santa Fe and on to Chihuahua. The caravan was suspected of carrying arms to the Mexicans and was pursued by American troops under Colonel Stephen Kearny. Wislizenus and his companions reached Chihuahua safely only to be attacked by an anti-American mob. They were sent under guard to a village in the Sierra Madres ninety miles west of Chihuahua, a village Wislizenus described as a "strange-looking, incomprehensible, awful place." Eventually he was rescued by American troops and joined them as a surgeon before returning to St. Louis. Engelmann had evidently briefed him on the need to botanize in out-of-the-way places, and he brought or sent back a remarkable collection of plants, some of which have rarely been found since.

Wislizenus's adventures attracted the attention of Senator Thomas Hart Benton, who saw to the publication of his *Tour Through Northern Mexico* as a government document. Included was Engelmann's account of the plants he had collected. While in Washington, Wislizenus met Lucy Crane, sister-in-law of diplomat George P. Marsh. Miss Crane could not then be persuaded to marry him, but two years later Wislizenus traveled to Constantinople, where Marsh was stationed, and convinced Miss Crane to become his wife.

Wislizenus settled in St. Louis with his family and was instrumental in helping to found the Missouri Historical Society and the Academy of Sciences of St. Louis. Engelmann was generous in naming plants for his good friend. The name *Wis-lizenia refracta* he applied to a conspicuous late-summer flower of the deep Southwest; the plants grow up to four feet tall, chiefly along roadsides and in dry streambeds. They belong to the caper family, but because of the resemblance of the leaves to those of a clover, they go by the name "jackass clover." Engel-mann also named a gentian, a giant barrel cactus, and other plants for Wislizenus. To these Asa Gray added *Dalea wislizeni,* a legume, and Swiss botanist Augustin De Candolle *Quercus wislizenii,* interior live oak. For a person who claimed to have little knowledge of natural history, he fared well.

Watson's Beardtongue

Sereno Watson (1826–1892), like his contemporary Wis-lizenus, had no initial background in natural history. In fact, he was forty-two before he joined the expedition of Clarence King in California and began to botanize. A graduate of Yale, Watson had tried teaching, medicine, banking, and farming. Failing in all of these, he set out in 1867 to try his luck with the King Expedition. King thought him a poor specimen: he was middle-aged, nearsighted, painfully shy, and in sorry physical condition. But King pitied him and took him on as a camp helper, without salary. About this time the botanist of the expedition, William Whitman Bailey, became ill and discouraged, while Watson showed skill in collecting plants and making excellent field notes. Bailey was happy to relinquish his post to Watson, who stayed on with King for several seasons and adapted to a rugged frontier life. During a survey of Great Salt Lake, he, Robert Ridgway, and two others capsized their boat and had to hang on to the hull for several hours in the cold, salty water before they were rescued. Watson was later invited to write up the botanical results of the expedition, a five-hundred-page, well-documented volume which he completed with the help of Daniel Cady Eaton and Asa Gray.

Publication of the volume ensured his reputation as a leading botanist, and in 1873 he was invited to Cambridge as curator of the Gray Herbarium. Here he prepared a two-volume *Botany of California*, updated Gray's *Manual of Botany*, and made many other contributions to his science. He did little further traveling aside from a trip to Guatemala that was aborted when he became ill. Gray named several plants for him, including *Galium watsoni*, a bedstraw, and *Penstemon watsoni*, a beardtongue. Not everyone would appreciate having a beardtongue named for him, since it would seem to imply that his beard had invaded his mouth. Watson is not likely to have been bothered by this implication; these are quite beautiful plants, the tubular blue flowers having throats tinged with red. Members of the large genus *Penstemon* have a sterile, flattened stamen that is "bearded" with long hairs, hence the name beardtongue. The genus *Watsonia* was, incidentally, named not for Sereno, but for the English botanist Sir William Watson.

Williamson's Sapsucker

Williamson's sapsucker is a bird of striking coloration, the male mostly black but streaked with white on the head, wings, and rump, and with a red throat and yellow breast. The female is equally distinctive, having a brown head and brown and white "ladder back." Like other sapsuckers, these birds make rows of small holes in the bark of trees and return to feed on the exuding sap and on the insects that are attracted to the sap.

Robert Stockton Williamson (1824–1882) was an army lieutenant in charge of the Pacific Railroad Survey in California and

Oregon. The surgeon on the survey, John Strong Newberry (1822–1892), took a male of an unusual woodpecker in 1857 and named it for his commanding officer. John Cassin had described an apparently very different sapsucker from the female sex six years earlier. When it was discovered that Williamson's and Cassin's sapsuckers were male and female of one species, Cassin's scientific name was retained, since it had priority. However, the common name Williamson's sapsucker is that accepted by the American Ornithologists' Union. Newberry served as surgeon-naturalist on several western expeditions. He was primarily a paleontologist, and after his military experiences he became a professor at Columbia University and published a major monograph, *The Palaeozoic Fishes of North America*.

Xantusia

*A*mong the least-known denizens of the baking deserts of the deep Southwest are the night lizards. No more than three to six inches long, these lizards hide among rocks and debris during the day and venture out at night to feed on termites and whatever other insects they can catch. Their nocturnal habits contrast with those of most lizards, which are creatures of the sun. Only a few intrepid herpetologists are familiar with these lizards, and no one has studied them in detail. When discovered during the day, they quickly look for a place of concealment and may even run up a person's pantleg in their haste to take shelter. Usually brownish with darker spots, these lizards turn pale in color when frightened and during nighttime forays. This stands in contrast to most lizards, which tend to be paler at higher temperatures and darker when it is cool, taking advantage of the fact that darker colors provide greater heat absorption. But night lizards are palest when it is dark and cool; perhaps in this way they can take advantage of what little radiation there is at night and so warm themselves at least a little during their nightly expeditions. Night lizards are strange in another way: rather than laying a clutch of eggs like most lizards, the females give birth to living young, only one or two, and the newborn lizards are nearly as large as the mother.

Night lizards possess enough structural peculiarities—for example, the lack of eyelids and the presence of large, rectangular plates on the underside—that they are placed in a genus of their own, *Xantusia*, and in a family apart from other lizards. There are four species, the best known being the desert night lizard of southern California and adjacent parts of Nevada and Arizona. It was discovered at Fort Tejon, near present-day Bakersfield, California, in 1857, and named for its discoverer, John Xántus (1825–1894), by Spencer Baird. The desert night lizard he called *Xantusia vigilis*, *vigilis* being the Latin equivalent of our word "vigilant," perhaps with reference to the fact that these lizards are especially hard to catch once they have been disturbed from their diurnal hiding places.

Xántus is one of the most enigmatic figures in the annals of natural history. Colorado naturalist–nature writer Ann Zwinger, who finds excitement in even the remotest corners of nature, and heroic qualities in even the most eccentric of naturalists, has been especially intrigued by Xántus. He was, she says, among other things, a "professional complainer, poseur, holder of grudges, irritant to almost everyone with whom he came in contact." But Baird prized him as a collector and wrote in 1858 that "[All] are unanimous in saying that there is but

Xantusia

one Xántus, unapproachable, and not to be equalled however much the second person may do." Both evaluations are fair.

Xántus was a native of Hungary, where he was trained in law. He enlisted in the cause of Hungarian independence but was taken prisoner and later forced to serve in the Austrian army. He escaped to Germany and to London, then in 1851 sailed for America. He landed in New York with seven dollars in his pocket but made a go of it for several years by teaching piano and languages and working as a clerk and even as a ditchdigger. In a series of letters to his family in Hungary, translated and edited by Theodore and Helen Benedek Schoenman in 1975, he told of his early American experiences with many embellishments and with (in the Schoenmans' words) "a pathetic tendency toward self-aggrandizement." For a time he lived in New Orleans, then for a time in a Hungarian settlement in Iowa. Desperate to find employment, "in a moment of utmost despair" he enlisted as a private in the army in 1855. He felt so humiliated by this step that he adopted an assumed name, Louis de Vesey, which he used off and on for the rest of his life. Army life was even more degrading than he had expected. "I had not the farthest idea [he wrote] that in the American army only the officers are considered *men*; and the others something like *last class negroes*."

Yet this was the beginning of his career as a naturalist. At his first post, Fort Riley, Kansas, he met Dr. William Alexander Hammond, an amateur ornithologist who was collecting specimens for Baird. Hammond taught Xántus how to collect and prepare specimens, and soon Xántus was sending material directly to Baird. This was the beginning of a long correspondence between Xántus and Baird, and the latter shortly used his considerable influence to have Xántus transferred to a relatively unexplored part of the country: Fort Tejon, California. Xántus was now a sergeant in the medical department, but his superiors looked askance at the amount of time he spent collecting specimens when he should have been performing his duties. Xántus chafed under authority, and when his pet grizzly bear cub ate the colonel's dog, conditions became strained indeed. Xántus complained that "here everybody is a gambler and drunkard, they sit day & night in whisky shops, or gambling holes; and instead of supporting me they ridicule my sport and throw every obstacle in my way."

Xántus was at Fort Tejon for slightly less than two years. Despite his problems, he shipped off many specimens to the Smithsonian, the Philadelphia Academy of Natural Sciences (which had elected him a life member), and the Hungarian National Museum. For some of these he was paid. Records of the Smithsonian show that he sent there 1,794 bird skins, 145 mammals, 229 jars of reptiles and fish, 107 bottles of insects, 14 bales of plants, and 17 packets of minerals. His many letters to Baird from Fort Tejon have been edited and annotated by Ann Zwinger; they are filled with complaints and esoteric details of his collecting, but they tell of a time when California condors were common and one could not "walk out half a mile, without meeting [a grizzly bear]." Xántus's adventures in California were written up for publication in Hungary and have been translated and edited by the Schoenmans as *Travels in Southern California* (1976). The book is good reading and illustrated with well-executed original drawings, but neither the text or the drawings can be completely trusted.

When he left Fort Tejon, Xántus was discharged from the army and was employed by the U.S. Coastal Survey at Cabo San Lucas, Baja California. Once again Baird had made these arrangements, and he reaped a considerable reward in specimens. Cabo San Lucas was even more poorly known to science than Fort Tejon, and there were many new species to be found. Xántus lived under primitive conditions and after twenty-eight months decided he had had enough. One of his letters, quoted

by Ann Zwinger in *A Desert Country Near the Sea* (1983), well
shows his despair.

> I am quite sick indeed of this place, every day seems a long
> year, and every one with the same monotonous desolation
> around me. . . . I am now of Gods grace nearly two years
> perched on this sandbeach, a laughing stock probably of the
> Pelicans & Turkey buzzards. . . .
>   There is not a blade of grass in the country, & not a green
> leaf. Sand, salt, trunks of shrubs, Rocks & the like everywhere
> and covered everywhere with bleached bones of catle mules
> horses etc., died by thousands lately of starvation.

Remarks by John Steinbeck and Edward Ricketts in their book
Sea of Cortez suggest that Xántus's life at the cape may have
been more enjoyable than he claimed.

> Speaking to the manager of the cannery at the Cape [wrote
> Steinbeck and Ricketts], we remarked on what a great man
> Xántus had been. Where another would have kept his tide
> charts and brooded . . . Xántus had collected animals widely
> and carefully. The manager said, "Oh, he was even better
> than that." Pointing to three little Indian children he said,
> "Those are Xántus's great-grandchildren," and he contin-
> ued, "In the town there is a large family of Xántuses, and a
> few miles back in the hills you'll find a whole tribe of them."
>   We wondered what modern biologist . . . would have the
> warmth and breadth, or even the fecundity for that matter,
> for all his activities. He at least was one who literally did pro-
> liferate in all directions.

In 1861 Xántus left for Hungary, but he failed to find a position
there and returned to the United States the following year.

Through his contact with Hammond, now U.S. surgeon general, he was appointed acting assistant surgeon general (though he had had no formal medical training!). But he soon resigned to accept a position, arranged for him by Baird, as U.S. consul in Manzanillo, Mexico. Here he was embroiled in controversy and was summarily dismissed after a few months.

Hammond had by this time been court-martialed and dismissed from the army, so he could no longer help Xántus. Hammond was a person of high standards and had come into conflict with Secretary of War Edwin Stanton, now best remembered for his disloyalty to President Lincoln. The court-martial found Hammond innocent of the trivial charges that had been brought against him, but Stanton insisted on a retrial that resulted in a guilty verdict. After his discharge from the army, Hammond had a successful career as a specialist in neurology at Columbia University and the City University of New York. By an act of Congress, in 1878, he was exonerated of the charges brought against him many years earlier and given the rank of brigadier general, retired.

John Xántus, too, despite his many discouraging problems, went on to a successful career. He returned once again to Hungary, where his two volumes of travels had made him well known. He became involved in the founding of the Zoological and Botanical Gardens in Budapest, and later a statue was erected to him at the entrance to the zoo, with a plaque identifying him as founder and first director. From 1869 to 1871 he traveled in East Asia and Indonesia, sending back specimens to the Gardens and the National Museum of Hungary. On his return he was made curator of ethnology at the museum, a position he held until his death.

Aside from his two books and his many letters, Xántus wrote two scientific papers, published by the Philadelphia Academy of Natural Sciences. In these he described Hammond's flycatcher (named for his first mentor), Cassin's vireo, and the spotted owl (now very much in the news). A great many species were described by others from his collections. Asa Gray described well over one hundred plants in a single paper, and over three hundred species of animals were named from his collecting. According to Ann Zwinger, nearly fifty species bear the name *xantusi*, *xantusii*, *xantii*, or *xantisiana*. Xántus's hummingbird, *Hylocharis xantusii*, is one of the most notable of these. Added to these are quite a number of species names based on his other name; for example, George Horn named one beetle *Cymatodera xanti* and another *Pachymerus veseyi*. It is as if he were two men, and perhaps indeed he did the work of two men during the thirteen years he spent in North America.

What, finally, can be said of this strange man? To the Schoenmans he was a person of heroic proportions, and Baird was lavish in his praise, though sorely tried at times. He was arrogant and often unable to get along with his superiors, though he seemed to enjoy his relationships with Native Americans. The records he kept with his specimens were often rather carelessly made, and his popular writings are full of invented stories of his escapades and discoveries, sometimes based on his reading of Frémont and other western explorers. He was surely one of the most intrepid collectors of all time, and Baird saw to it that his energies were expended in places then little known to science. Naturalists are, after all, remembered for their accomplishments and not, fortunately, for their charm and good manners.

Yarrow's Scaly Lizard

*D*inosaurs have always fired our imaginations and nowadays can be purchased at any novelty store: stuffed, inflatable, made of plastic, paper, or metal. Children are likely to know all about tyrannosaurus and triceratops, and paleontologists are forever coming up with new kinds as well as new theories about their demise. But who knows the ways of those diminutive cousins of the dinosaurs, the lizards that are still very much with us, scampering about the rocks in the sun? There are about 125 species in America north of Mexico, and all go about their ways unobserved except by a few naturalists who don't know enough to come in out of the sun.

The term "scaly lizard" at first seems odd, since all lizards have scales, but in this group (genus *Sceloporus*) the scales are unusually coarse and bristly. These are squat, short-legged lizards that spend a good deal of time on rocks or trees. Before split-rail fences were replaced by barbed wire, they often liked to perch on them, so they came to be called "fence lizards." Yarrow's scaly lizard (*Sceloporus jarrovii*) occurs in the mountains of New Mexico, Arizona, and northern Mexico, from five thousand feet elevation up to about ten thousand feet. The scientific name requires a bit of explanation. The species was

Yarrow's Scaly Lizard

named for Henry Crècy Yarrow, but since there is no *y* or *w* in Latin, the nearest equivalents, *j* and *v*, had to be used. *Sceloporus*, by the way, is Greek for leg-pore, an appropriate name since the males of these lizards have glandular pores on their legs. During the mating season, the pores produce a substance that is apparently attractive to females of their species.

Yarrow's scaly lizard is about three and a half inches long, not counting a tail that is slightly longer than that. It is rather drab in color, but the males have a bright spot in the center of each scale and a splash of blue on their throat and sides; both sexes have black collar markings. Like many lizards (but unlike *Xantusia*) they tend to be paler when it is hot and darker when it is cool. Experiments have shown that the heat gain from light to dark is 7.4 percent, an amount that may be very important to a cold-blooded animal that ranges into high altitudes.

During the mating season, male scaly lizards challenge other males that enter their territory. They do so by flattening their body and extending a throat pouch, exposing the blue coloration. If the intruding male does not flee, a fight may follow.

The blue color apparently plays no important role in courtship. G. K. Noble of the American Museum of Natural History studied the behavior of fence lizards experimentally, back in 1934, and concluded that "the adornment of the males is not wedding finery but a gladiator's vestment." He felt that the lizards failed to confirm Darwin's ideas on sexual selection. Nowadays we would say that the display of blue coloration functions in intrasexual selection (rivalry within one sex) rather than intersexual selection (that is, choice of a partner).

Noble found that when a female is painted blue beneath, she is treated as if she were a male. Otherwise a female is approached with a series of head bobs, and if receptive the female will switch her tail from side to side. Don Hunsaker of San Diego State College provided several females with "chastity belts" by taping over their genital apertures. Females so treated would switch their tails and climb over and under males to attract attention. This activity extended over two or three weeks if the female was not allowed to satisfy her sexual drive. It should be added that Hunsaker kept the lizards in laboratory cages and, during the winter, injected them with sex hormones to induce mating.

Hunsaker kept several species of scaly lizards in his laboratory, and it appeared that the head bobs differed in pattern in each species. In order to study this behavior in more detail, he developed a "head-bobbing machine," consisting of a small motor with a rod attached and a thread suspended from its end, rather like a small fishpole and line. A plastic model of a lizard was then attached to the end of the line so that it could be "bobbed" in various patterns so as to determine the response of lizards to it. In this way he established that, in fact, each species would respond only to the pattern of bobs characteristic of that species. Yarrow's, for example, has two short bobs, a longer one, and another short one—as if it had its own signature in Morse code. Within its normal range, another species of *Sceloporus* is common, *torquatus*, and the two may sometimes be found on the same rock. But *torquatus* has a single, more intense head bob. Since the various species all look much alike, presumably they tell each other apart in nature, at least in part, by the pattern of bobs with which they greet one another. Thus no time or effort is wasted in mating with an alien species.

Henry Crècy Yarrow (1840–1929), for whom this species was named, was born in Philadelphia and trained as a physician at the University of Pennsylvania. Following service in the Civil War, he was posted to various army bases, at one of which he became acquainted with Elliott Coues. The two studied the fauna of Fort Mason, North Carolina, and published a report on it in 1878; that was evidently Yarrow's introduction to natural history. Spencer Baird was impressed by the young surgeon and recommended him to the Geographical Survey of the West headed by Lieutenant George M. Wheeler. In his report on the ornithology of the survey, Yarrow described the natural history of the water ouzel (or dipper), Gambel's quail, and several other birds. After his return from the survey, he became associated with George Washington University as a professor of dermatology. He had been interested in reptiles for some time, and he was eventually appointed curator of reptiles at the U.S. National Museum. He was also associated for a time with the U.S. Fish Commission. By all accounts Yarrow was a congenial person, one of the founders of Washington's Cosmos Club and part of an active group of naturalists in that city in the late 1800s and early 1900s.

Edward Drinker Cope (1840–1897), who described Yarrow's scaly lizard, was born in the same year and in the same city as Yarrow. The two later spent time together in the field as members of the Wheeler survey. Cope came from a family of wealthy Quaker farmers, and his father wanted him to become a farmer to the extent that he bought a farm for him. But when he was no more than eight years old, Edward began visiting the

Philadelphia Academy of Natural Sciences and describing in his journal what he saw. When he was nineteen he went to Washington to study reptiles with Spencer Baird; his first scientific paper was published that same year. In 1861 the Philadelphia Academy of Natural Sciences elected him to membership, and a few years later he was appointed curator there (without pay). Cope then decided to sell his farm, invest the money, and launch himself on a career as a naturalist.

Cope's first interest was in living reptiles, and he retained this interest throughout his life. A leading journal of herpetology, *Copeia*, is named for him. But his home was near fossil beds that contained the remains of ancient marine reptiles, plesiosaurs and mosasaurs, and he began to study them. One of his field trips to the fossil beds was with a friend from Yale, Othniel Charles Marsh. In 1871 Cope decided to head for similar beds in Kansas, which had earlier been studied by Marsh. A year later he followed in the wake of Marsh again, this time joining the Hayden survey in the Bridger Basin of Wyoming. Since Cope and Marsh were working the same fossil beds and both were rushing into print with their discoveries, it was inevitable that they would sometimes apply their own names to the same fossils, resulting in a squabble over priority. This was the beginning of a feud between these former friends.

In 1874 Cope joined the Wheeler survey in the San Juan Basin of northern New Mexico. Here he found Yarrow in charge of a group that was mapping the geology of the area. But when Yarrow wanted to move on, Cope insisted they stay, as he was finding good fossil collecting. This led to an altercation that was finally settled in Cope's favor by the commanding officer in Santa Fe. Cope was next off to the badlands along the Judith River of Montana (named by William Clark for his wife-to-be). Nearly a ton of bones was shipped down the Missouri and on to Philadelphia.

Cope and Marsh locked horns again in 1877, when Arthur Lakes, a schoolteacher in Golden, Colorado, found bones of a "super-dinosaur" and sent some of them to Marsh and some to Cope. Marsh published first, reporting that the dinosaur must have been fifty to sixty feet long, the largest land animal yet known. Cope was not to be outdone and hired a party to dig in similar beds near Canon City, Colorado. His money was now seriously depleted, and he invested what was left in several mining properties. But by 1886 his mines had failed, and three years later he accepted a professorship at the University of Pennsylvania, which he held until his death. Cope and Marsh continued to argue over the ownership of specimens and the priority of names.

Cope's bitterness toward Marsh came to a head in an article in the New York *Herald*, which I quote only in small part. "Professor Marsh's reputation for veracity among his colleagues is very slight. . . . Those who know him best say—and I concur in the opinion—that he has never been known to tell the truth when a falsehood would serve the purpose as well."

Marsh responded by pointing out that one of Cope's most remarkable finds was in fact remarkable only because Cope had it backward; what he had called the tail was actually the head.

Major John Wesley Powell, with whom Marsh had worked, joined the fray: "The Professor [Cope] has done much valuable work for science. . . . If his infirmities of character could be corrected by advancing age, if he could be made to realize that the enemy which he sees forever haunting him as a ghost is himself . . . he could yet do great work for science."

In 1895 Cope sold his collections to the American Museum of Natural History in New York. He died two years later in Philadelphia. At his funeral, the mourners gathered around his coffin "amidst the fossil bones, with a pet live tortoise and a Gila monster moving stealthily about the room." (The quote is from *The Bone Hunters*, in which Url Lanham tells of the Cope-Marsh dispute in much greater detail.) The coffin, in fact, was empty. Cope had seen that his skeleton was placed in a museum and his brain pickled in formalin. He was a naturalist even in death.

Zenaida

*D*oves have a rather hard time of it throughout the world. Unfortunately they are good to eat. The last passenger pigeon died in 1914; the white-winged dove of our southwestern states is becoming less abundant; worldwide, half a dozen other doves and pigeons are listed as threatened or endangered. Fifty million mourning doves (*Zenaida macroura*) were killed by hunters in the 1988–1989 season. Nevertheless these familiar birds seem to be doing pretty well. Except in forests, there is hardly anyplace in temperate North America where one cannot hear their gentle *cooWOO, coo, coo, coo*, as if to say, Here we are, well but in mourning to think that all of this must pass. The nests are such flimsy structures, containing only two eggs, that one wonders that their populations are able to keep ahead of the hunters, who need several to make a decent meal. They breed twice, sometimes three or four times a year, a fact that at least partly accounts for their continued success.

Mourning doves remain paired for the season. A courting male approaches his mate with his head low and his body raised on tiptoes; he then inflates his neck to display a pair of iridescent patches while cooing softly. At times he flies upward with noisy wingbeats, then glides down in a broad circle with his wings spread. Before mating, the male offers his open bill to the female, who inserts her bill into his, often with an exchange of food. The term "billing and cooing," sometimes applied to human behavior, is adopted from remotely similar behavior in doves.

Male and female work together to build the nest and feed the young. As with other doves, the young are fed at first with "pigeon milk," produced by both males and females from glands in their crops. After a few days the young are fed with small seeds. Once they have left the nest, doves feed on the ground, not by scratching with their feet, as chickens do, but by flicking their bill from side to side. Seeds of grasses and weeds make up most of their diet. In agricultural areas, corn, wheat, and other

Zenaida

grains are consumed, but only seeds lying on the ground after harvest; the doves pose no threat to standing crops. Oddly, they have a taste for snails. Perhaps they require minerals from the snails or the females calcium for eggshell production.

When disturbed, mourning doves fly off on whistling wings. They require water, and travelers in desert areas often watch their flight in order to find water holes. The subdued, grayish plumage of mourning doves suits well their quiet and unobtrusive behavior. The long, white-fringed tails and the iridescent spots on the neck are their only adornments. They are good birds to have around, birds of peace as legend tells us.

There are nearly three hundred species of doves occurring throughout the world, even on many isolated tropical islands. Larger species are usually called pigeons; some of the tropical species are large indeed, almost hen-turkey size. Among the oddest are the crowned pigeons of New Guinea, which have a fan-shaped crest composed of lacelike feathers. Still odder is the tooth-billed pigeon of Samoa, a robust bird with a stout, hooked bill resembling that of a dodo. The now extinct, giant flightless dodo of Mauritius was, in fact, a pigeon.

It is curious that one of the many species of doves has become adjusted to living closely with people: the rock dove, or domestic pigeon. These birds once lived on rocky ledges in North Africa, but they now occur in cities throughout the world, living on the ledges of buildings and feeding on whatever scraps they can find. Pigeon fanciers have bred a great many ornate varieties, and of course homing pigeons have found a place in history. Much of what we know about bird navigation is based on experiments with pigeons. They are so commonplace that we tend to ignore them, though in fact there is much to be said for them. As human populations continue to grow, it is species such as the common pigeon that will flourish while others disappear.

It was Linnaeus who chose the species name *macroura* (Greek for long-tail) for the mourning dove, but he grouped all doves in one genus (*Columba*) despite their diversity. The uniqueness of this and several other, similar species was recognized by a person who at first seems improbable: a nephew of Napoléon, Charles Lucien Bonaparte (1803–1857). Charles had no taste for warlike pursuits, and not until late in life did he show a penchant for politics. He became intrigued by birds as a child, and after he married he named the genus of the mourning dove after his wife Zenaide. Zenaide bore him eight children, but it was evidently not the happiest of marriages, for Charles and Zenaide later separated. Zenaide died in 1854, but thanks to her husband her name remains attached to a genus of handsome birds that are often thought of as symbols of loving attachment.

When Charles Bonaparte was born in 1803, Napoléon was about to declare himself emperor of France, and he was displeased that his brother Lucien had married a commoner rather than a person of royal blood. Lucien and his family fled to Italy, but even there they could not escape Napoléon's influence. They went on to England, where their son Charles first developed an interest in natural history. Back in Italy after the end of the Napoleonic wars, nineteen-year-old Charles married his cousin Zenaide, the daughter of Joseph Bonaparte, who had been king of Naples and Spain and now lived near Philadelphia. Napoléon was by then on St. Helena, and he died before the wedding took place.

The newly married couple moved to Philadelphia, where Charles began a study of American birds. He became a friend of Audubon, Titian Peale, Thomas Say, William Cooper, and other naturalists. It was Say who assisted him in producing a supplement to Alexander Wilson's *American Ornithology*, in which Bonaparte described Say's phoebe and Cooper's hawk. William Cooper (1798–1864), for whom the hawk was named, was a prominent New York naturalist and one of the founders of the Lyceum of Natural History. He described several birds, including the evening grosbeak. (It was his son, James Graham

Cooper, who wrote *The Ornithology of California* and for whom the Cooper Ornithological Society was named.)

Charles Bonaparte bore a strong resemblance to his uncle Napoléon. At a meeting of the Philadelphia Academy of Natural Sciences in 1825, a visitor, Dr. Edmund Porter, described him as "a little set, blackeyed fellow, quite talkative, and withal an interesting and companionable fellow." And he added: "To a novice it seems curious, that men of the first intelligence should pay so much attention to web-footed gentry with wings."

Bonaparte sent Titian Peale to Florida to collect for him, then the following year produced the first volume of his treatise on American birds. Although said to be an extension of Wilson's book, it was really very different, the product not of a field naturalist but of one who was much more at home in libraries and museums. Bonaparte next planned to review the classification of birds of the world, and he felt that only in Europe could this be accomplished. So in 1826 he traveled to England, Germany, and Italy to study museum specimens.

In London he visited once again with Audubon, who remarked on the meeting in his *Journal.* "I cannot tell you how surprised I was when at Charles's lodgings to hear his manservant call him 'your Royal Highness.' I thought this ridiculous in the extreme, and cannot conceive how good Charles can bear it. . . ."

In fact Bonaparte cherished his title, Prince of Canino and Musignano. After a brief visit to Philadelphia in 1827 to see to the publication of further volumes of his book on American birds, he left America for good in 1828, settling in Rome. He invited his old friend William Swainson to join him in writing his *Conspectus* of the birds of the world, but Swainson declined. Swainson nevertheless characterized Bonaparte as "destined by Nature to confer unperishable benefits on this noble science."

Bonaparte became involved in the cause of Italian independence, but in 1848 Italy fell to the forces of his cousin Louis

Napoléon, president of France. Once again he fled, first to England and later, after being pardoned by his cousin, to France. Here he finally completed the first part of his *Conspectus*, a monumental publication in which many new genera and species were described. He continued to work steadily on the remaining parts despite declining health. A visitor found him writing in his bath a short time before his death in 1857. There have been many tributes to him, including the following from Erwin Stresemann's *Ornithology from Aristotle to the Present*.

> None of those who encountered this great ornithologist could resist the spell of his personality, which combined a fiery Roman temperament with intelligence and sound judgment, and a cheerful disposition. . . . With absolute certainty, this remarkable man found the right way into the profoundest depths of science, because of his all-penetrating perception and his almost excessively intense activity.

One of those who had reservations about Bonaparte was the ever-critical Elliott Coues. "In his later years, Bonaparte simply played chess with birds, with himself as king: *le roi s'amuse!* Scheme followed scheme, tableau tableau, conspectus conspectus, with perpetual changes, incessant coining of new names, often in mere sport—it was nothing but turning a kaleidoscope. It may have been fun for him, but it was death to the subject."

Although Zenaide's name is preserved as the name of a genus of doves, the species of gull named for Charles, *Larus bonapartii*, is now called *Larus philadelphia*, since that is an earlier name for the species, and priority prevails. Nevertheless the accepted common name is Bonaparte's gull. These small, black-capped gulls are often seen swirling in flocks over the breaking waves on both coasts. They will surely do as a reminder of Bonaparte's remarkable career.

Women in Natural History
of the Nineteenth and Early
Twentieth Centuries

When Charles Bonaparte named a genus of doves for his wife Zenaide, he was one of many who named genera or species after women they admired. I have mentioned several others: Lucy's warbler, named for Baird's daughter; Virginia's warbler, dedicated to Anderson's wife; Grace's warbler, named by Baird for Coues's sister; Anna's hummingbird, named by French ornithologist Lesson for Anna, Duchess of Rivoli; Queen Alexandra's sulphur butterfly, named for the wife of Edward VII; and Edith's copper butterfly, named for Theodore Mead's wife. To this may be added the blue-throated hummingbird, *Lampornis clemenciae*, named by Lesson for his wife.

There has been some confusion over the naming of the blackburnian warbler. Evidently the British naturalist Thomas Pennant acquired a specimen from a museum maintained by a wealthy patroness of science, Anna Blackburne (1726–1793). He gladly called it the blackburnian warbler, since he was indebted to Miss Blackburne for many specimens he included in his book *Arctic Zoology*. But he did not propose a scientific name, and it remained for the German naturalist J. F. Gmelin, a few years later, to Latinize the name to *blackburniae*. But in fact another researcher had found the same species in its winter home in South America and some years earlier had proposed the name *fusca*. *Fusca* is the Latin word for "dusky" and is particularly inappropriate for one of America's most beautiful warblers. Priority must, as usual, prevail, and the species is now called *Dendroica fusca*. Fortunately blackburnian warbler remains the vernacular name.

Anna Blackburne is said to have been a handsome, statuesque woman who never married but devoted her life to her gardens and museum. She did no research herself, but she was often helpful to those who did. Linnaeus said of her: "She is a true naturalist whose esteem I covet." The plant genus *Blackburnia* was also named for her. When the Danish entomologist J. C. Fabricius visited her museum, he found an undescribed dung beetle which he named *Scarabaeus blackburni*. His use of the masculine ending *i*, rather than the feminine ending *ae*, indicated that he had named it not for Anna, but for her brother Ashton, who was a major contributor to her museum.

Clearly naturalists liked to have particularly charming birds or butterflies (but no bats, rats, or dung beetles) named for their wives, daughters, sisters, or other attractive women. Otherwise eighteenth and nineteenth century women were expected to remain in the background and bear children. To be sure, some wives were intimately involved in their husbands' research. Lucy Say, for example, drew excellent sketches of shells, and after her husband's death was elected to the Philadelphia Academy of Natural Sciences. Elizabeth Merriam went on long field trips with her husband, often enduring rough weather and primitive living conditions. Elizabeth Agassiz presented some of her husband's research in popular form in her book *Seaside Studies in Natural History*. But these women were merely participants in their husbands' careers. Opportunities for women to become well educated were limited until well into the nineteenth century, and ideas of women's liberation were only beginning to stir.

Jane Colden (1724–1766), sometimes regarded as America's pioneer woman scientist, was trained by her father, amateur botanist Cadwallader Colden, of Newburgh, New York, as his assistant. Jane classified over three hundred plants of the Hudson Valley and sketched many of them. Joseph Kastner tells the story that it was proposed to Linnaeus that he name a plant for her, whereupon her aunt remarked, "What, name a weed after a Christian woman!" But Jane Colden published very little, and following her marriage in 1759 she fell into a role more typical of women of her day.

In the nineteenth century there were several women who deserve to be called naturalists. Few were able to establish professional careers, with the notable exception of Maria Mitchell (1818–1889), who discovered a comet and went on to a long career as a professor of astronomy at Vassar College. Almira Hart Lincoln was a student in one of Amos Eaton's classes and later wrote *Familiar Lectures in Botany* (1829), which became a standard reference in female seminaries current at the time. It went through ten editions and sold over 275,000 copies. Mrs. Lincoln noted that "animals cannot be dissected and examined without painful emotions. But the vegetable world offers a boundless field of inquiry, which may be explored with the most pure and delightful emotions."

Genevieve Estelle Jones (1847–1879) wrote *Illustrations of the Nests and Eggs of Birds of Ohio*, completed by friends after her death at the age of thirty-two. Elliott Coues reviewed it favorably in the pages of *The Auk*, the publication of the American Ornithologists' Union. Graceanna Lewis (1821–1912) was a protégée of John Cassin, curator of birds at the Philadelphia Academy of Natural Sciences; she was later elected to the Academy. Spencer Baird encouraged her and helped her to obtain a microscope for her careful studies of bird feathers. Although she lectured widely and wrote several articles, she was unable to obtain a position at a university or museum, even though Baird considered her "a very remarkable woman for her attainments and accomplishments in natural history."

Perhaps the most interesting of the nineteenth century woman naturalists was Martha Maxwell (1831–1881). Oberlin College had been opened as a coeducational institution in 1833, and a few years later Martha traveled by wagon from her home in Wisconsin to enroll. Tuition was twelve dollars a year, but even so she was too poor to remain more than a year. A few years later she married James Maxwell, a widower with six children who was shortly lured to the goldfields of Colorado. Here Martha, who had been interested in animals from childhood, met a German prospector who taught her taxidermy. In time she bought a gun, practiced marksmanship, and traveled into the mountains to collect specimens. Her Rocky Mountain Museum, first in Boulder and then in Denver, attracted attention for the variety of animals expertly mounted and placed in seminatural settings unique for their time. Soon she was corresponding with Spencer Baird, Elliott Coues, and Robert Ridgway.

In 1876 Martha Maxwell was invited to prepare an exhibit for the Centennial Exposition in Philadelphia. Her display proved to be one of the hits of the exposition, and she later moved it to Washington for a time. Coues prepared a list of the mammals she had collected in Colorado; it included a black-footed ferret, which Coues himself had never seen in the wild. Ridgway similarly prepared a list of the birds, including several species seldom reported from Colorado, as well as a new subspecies of screech owl which he called *maxwellae*.

The exhibit was not profitable, however, and Mrs. Maxwell was unable to persuade any museum to purchase it. When she died in 1881, at the age of fifty, the collection went to her daughter, who entrusted it to an entrepreneur in Saratoga Springs, New York. He exhibited it for a time, then put it in storage, where it gradually deteriorated. Martha Maxwell's techniques of taxidermy and of habitat groups were well in advance of her time, and she was a keen student of nature despite her failure to break into the ranks of more established scientists.

While Martha Maxwell's exhibits were being displayed at the Centennial Exposition, a young woman from central New York was studying at Cornell University, which had been founded a few years earlier as a coeducational institution. In 1878 she married her professor, entomologist John Henry Comstock. Anna Botsford Comstock (1854–1930) was a well-trained naturalist and a proficient illustrator of her own and her husband's books. In 1895 the Comstocks collaborated on the first of several editions of *A Manual for the Study of Insects.* In 1898 Anna was appointed assistant professor of natural science, but because of objections by the board of trustees she was soon downgraded to lecturer. (She was once again elevated to assistant professor in 1913, and received a full professorship in 1919, shortly before she retired.) Anna Botsford Comstock's *Handbook of Nature Study,* first published in 1911, has gone through twenty-four editions and is still in print; it has been translated into eight languages. It was designed primarily for schoolteachers but contains a vast amount of information most of us have forgotten, if we ever knew it. Mrs. Comstock believed that children should be introduced to their natural environment at an early age and that teachers should be qualified to teach them—something we seem to have forgotten in these times of widespread environmental deterioration.

Elizabeth Britton (1858–1934), like Anna Comstock, worked in close association with her husband, who was Nathaniel Lord Britton, director of the New York Botanical Garden (which he and Elizabeth helped to found) and coauthor of Britton and Brown's *Illustrated Flora.* Elizabeth Britton became a world authority on mosses, with a publication record of over three hundred titles.

Cornelia Clapp (1849–1934), on the other hand, followed an independent career as a professor at Mount Holyoke College, spending her summers at Louis Agassiz's laboratory on Penikese Island and later at the Marine Biological Laboratory at Wood's Hole (where she became a trustee). She held two Ph.D. degrees and an honorary Sc.D. and was a brilliant and influential teacher. Her major publication dealt with the sensory organs of the toadfish, a subject that hardly conforms to the stereotype of womanly devotion to delicate and attractive subjects.

While Anna Comstock, Elizabeth Britton, and Cornelia Clapp were having successful careers in the eastern states, Alice Eastwood (1859–1953) was serving as curator of botany at the California Academy of Sciences in San Francisco and adding in important ways to knowledge of the rich California flora. She was brought up in poverty in Colorado and had the simplest of educations, but one of her teachers sensed her interest in plants and gave her a copy of Asa Gray's *Manual of Botany*. Alice herself became a schoolteacher and spent her summers roaming the mountains for plants.

In 1892 Alice Eastwood was offered a curatorship at the California Academy of Sciences. She did much to build up the herbarium, but nearly all was lost in the earthquake of 1906. After a period of study in Europe, she returned to a rebuilt Academy and remained there for the rest of her long life. When she was ninety-one, she traveled to the International Botanical Congress in Sweden, where she had the thrill of being allowed to sit in the chair of the great Linnaeus. Alice Eastwood published over three hundred articles, and the herbarium she had gathered at the California Academy of Sciences numbered over 300,000 specimens. She was especially attracted to plants of the California chaparral and described one of its most characteristic shrubs, now called Eastwood manzanita. Another shrub of arid lands in California is *East-*

woodia elegans, a yellow-flowered member of the daisy family that is the only member of its genus. Either will do as a memorial to a woman whose love for plants carried her from poverty to the top of her profession.

There were other woman naturalists in the nineteenth and early twentieth centuries, several of them discussed by Marcia Myers Bonta in her book *Women in the Field: America's Pioneering Women Naturalists.* But not until well into the twentieth century were more than a few women employed in senior positions in museums or universities. Now, fortunately, it is possible to think of many women who have made major contributions to natural history. Anthropologist Louis Leakey maintained that women were more observant than men and often developed a closer rapport with animals. Three of his protégées, Jane Goodall, Dian Fossey, and Birute Galdikas, all students of apes in their native habitats, did much to support his belief. Margaret Rossiter's book *Women Scientists in America: Struggles and Strategies to 1940* provides an excellent source of information for the first four decades of the twentieth century.

In 1971 the publication *American Men of Science,* founded in 1906, was finally retitled *American Men and Women of Science.* Even the National Academy of Sciences, which is made up mostly of professors from self-styled "prestigious" universities who each year elect more of their colleagues, from time to time elects a woman. As late as 1991, women academy members were pleading for greater participation in the activities of the National Academy and the National Research Council. Nevertheless, when the history of natural science in the twentieth century is written, it will reflect the many important contributions and increasingly significant role of women in the natural sciences.

Naturalists, Then and Now

$\mathcal{T}$he era of the pioneer naturalist is past. Most North American birds, mammals, and plants are now reasonably well known, and we are now more concerned about losing them than finding them. Yet we owe a great deal to the men (and more rarely, women) who devoted their lives to roaming the fields and forests in order to make known the richness of our continent. They were a curious lot, and we shall not see their likes again.

They were often willing to endure incredible hardships for the sake of a few new plants or birds. Woodhouse was bitten by a rattlesnake and pierced by an Indian arrow; Douglas was trampled by a bull; Gambel died of typhoid fever after crossing the Sierra in winter; Lawson and Gunnison were tortured and killed by Native Americans; and Gregg died in California either from starvation or from falling from his horse while ill. Franklin and Richardson nearly perished of cold and starvation in the Far North, and Franklin later died there. Drummond died a mysterious death in Cuba, Bullock somewhere in South America, Steller in Siberia. Townsend poisoned himself with arsenic, a constituent of the preservative he used for bird and mammal skins.

Clearly these men were strongly motivated, almost beyond reason. Rafinesque's paean to the naturalist's life I have quoted earlier. In a brief autobiography, Nuttall wrote of his impressions on arriving in America:

> All was new!—and life . . . was then full of hope and enthusiasm. The forests apparently unbroken, in their primeval solitude and repose spread themselves on either side as we passed placidly along. The extending vista of dark Pines gave an air of deep sadness to the wilderness. . . .
>
> Scenes like these have little attraction for ordinary life. But to the naturalist it is far otherwise; privations to him are cheaply purchased if he may roam over the wild domain of primeval nature. . . .
>
> How often have I realized the poet's buoyant hopes amid these solitary rambles through interminable forests! For thousands of miles my chief converse has been in the wilderness with the spontaneous productions of nature; and the study of these objects and their contemplation has been to me a source of constant delight.

And here is Townsend, following the Platte River through Nebraska in May 1834:

> In the morning, [we] were up before the dawn, strolling through the umbrageous forest, inhaling the fresh, bracing air, and making the echoes ring with the report of our gun, as the lovely tenants of the grove flew by dozens before us. I think I never before saw so great a variety of birds within the same space. All were beautiful, and many of them quite new to me; and after we had spent an hour amongst them, and my game bag was teeming with its precious freight, I was still loath to leave the place, lest I should not have procured specimens of the whole.

None but a naturalist can appreciate a naturalist's feelings—his delight amounting to ecstacy—when a specimen
such as he has never before seen, meets his eye. . . .

This quote serves to remind us that pioneer naturalists were
sometimes ruthless by our standards. A gun was often an
important part of their equipment, and they showed little
restraint in using it to collect even the rarest and most enchant-
ing of birds. Earlier I quoted Elliott Coues on his encounter
with Atlantic puffins. Here is Edgar Mearns, after locating a
nest of the rare LeConte's thrasher in the Arizona desert:

After examining the nest, I concealed myself under a neigh-
boring mesquite. . . . At length, when I was almost roasted,
[the female] flew into the mesquite and almost immediately
took her place upon the eggs. A chirping call from me quickly
brought her to the top of the bush, where I shot her. With
the male the case was different. It required a chase of an hour
to secure him. . . . At length I winged him at long range
when flying, and then had an exciting chase upon the
ground, shooting him as I ran.

Mammal hunters often carried a string of traps, fish hunters
seines, entomologists their nets, botanists vascula that they
loved to fill with the rarest of plants. Most were adept at
preparing mammal and bird skins, or pressing plants. All kept
field notes, with varying degrees of care and thoroughness.
These notes often included data on habitat, feeding and nesting behavior, or nests; or in the case of plants, season of flowering or setting seed, soil type, and so forth. Over time, as data
accumulated, each species became more than a name—a population occurring within a particular area and interacting with
other species and with the environment in its own characteristic way.

There is no question that naturalists were regarded as oddi-
ties by settlers and by trappers, Native Americans, and others
they met on their travels. The voyageurs who accompanied
Nuttall on his trip up the Missouri called him "le fou" and
laughed at his absentminded dedication to collecting plants.
When Kansas Indians saw Say collecting beetles in their midst,
they accepted him as a medicine man. Wilson played Scottish
airs on his flute as he roamed the wilds, and Audubon carried
his fiddle. Palmer got in trouble when he tried to measure the
breasts of Indian women, and Townsend was caught trying to
steal a Native American corpse. Nelson collected Eskimo skulls,
according to John Muir, "like a boy gathering pumpkins." Xántus was an "irritant to almost everyone with whom he came in
contact." Rafinesque was so naive that he formally named and
described several fish that Audubon had invented. Girault
spent time in an asylum and lived on the fringe of madness for
much of his life. A strange lot, indeed.

The starry-eyed naturalist, laden with vascula and brandish-
ing a butterfly net, became a stock figure in literature. In James
Fenimore Cooper's novel *The Prairie* he appeared as Dr. Battius,
who roamed the plains and annoyed his companions by apply-
ing Latin names to everything in sight. Once, in the dim light
of dawn, he heralded his own donkey as a new species. On
another occasion, he paused in the midst of a skirmish to collect an unknown plant, forgetting "everything but the glory of
being the first to give this jewel to the catalogues of science."
Elsewhere Battius speaks of *Ursus horribilis* (the grizzly,
described by Ord from specimens taken by Lewis and Clark)
and of *Canis latrans* (the coyote, described by Say from the
Long Expedition). *The Prairie* was written in 1827, shortly after
Cooper had read the narratives of the Lewis and Clark and the
Long expeditions. He had never been west, and his evocations
of western scenes and of Native Americans, trappers, and settlers were all derived from his reading. Dr. Battius is a carica-

ture, but one sees in him a bit of Rafinesque, and of Nuttall, Say, James, and Wilson.

The rapid growth of descriptive natural history in the nineteenth century owed much to the Linnaean revolution of the mid–eighteenth century. With a simplified system of naming plants and animals that allowed them to take credit for their discoveries, naturalists took advantage of every opportunity to add to the catalog of nature. It was the time of the growth of the country's great natural history museums: the Academy of Natural Sciences of Philadelphia, the U.S. National Museum, the American Museum in New York, the Museum of Comparative Zoology and the Gray Herbarium at Harvard, Pittsburgh's Carnegie Museum, Chicago's Field Museum, the California Academy of Sciences, and still others.

It was the Darwinian revolution of the mid–nineteenth century that, over time, expanded the vision of naturalists from the discovery and description of new plants and animals to broader issues. Darwin wrote many books and articles besides *The Origin of Species*. He studied the pollination of plants and the actions of earthworms in moving soil—two of several topics we would now call *ecology*, although the word had not been invented in Darwin's time. His studies of animal communication did much to lay the foundations of the science of animal behavior. The science of genetics, too, rose partly as a consequence of Darwin's emphasis on the importance of variation in populations—and his failure to understand its basis. Meanwhile there was impetus for descriptive naturalists to become systematists, not only concerned with nature's diversity but also with the adaptiveness of organisms to their environment and to their origin and evolution. By the beginning of the twentieth century natural history had exploded into several disciplines, the practitioners of which preferred to call themselves biologists. Indeed, the word *naturalist* was often used in a derogatory sense, usually prefixed with the word *old-fashioned*. Is it true that the naturalist no longer has a role to play in our technological society?

A case can be made that there has never been a greater need for naturalists. The study of living things and life processes has now become so fragmented that biologists often do not communicate with one another, to say nothing of communicating with nonbiologists. Each discipline has developed its own techniques and terminology, and it is all too easy to become lost in a quagmire of obscure verbiage. Who is there to attempt an overview, to speak of the unity of nature, the interrelationships of all its parts, if not the naturalist?

Further, there is no one better qualified than the naturalist to point out the importance of each species as a unique combination of genes that has persisted for a long time and has evolved to play a particular role in our environment. The Endangered Species Act of 1973 is a constitution for naturalists; efforts to expand it and ultimately to extend it worldwide can occupy naturalists for many years to come. Meanwhile, naturalists must still fill their traditional role as discoverers and describers of life on earth.

The catalog of nature may be less than half complete—some say as little as 15 percent complete. The majority of undescribed species are small, to be sure: bacteria, fungi, nematodes, insects, and so forth. But smallness does not imply unimportance; rather the opposite—some of these organisms are the very underpinnings of life on earth. In a recent article in the journal *Science*, Peter Raven, director of the Missouri Botanical Garden (and thus an intellectual descendant of George Engelmann), and Edward Wilson, a professor at Harvard's Museum of Comparative Zoology (founded by Louis Agassiz), have pointed out that within the last decade botanists have described three wholly new families of plants from the American tropics, while zoologists have described eleven of the eighty known species of cetaceans (whales and dolphins) in this century, the most recent in 1991. Clearly not all recent discoveries are in the minuscule.

We ought to be interested in every species simply because it is there; in a sense each one is its own unique miracle. But if reasons are needed to learn of species—and to cherish them—there are reasons aplenty. About twenty kinds of plants now supply 90 percent of the world's food supply; these are grown in monocultures highly susceptible to disease and to insects that have developed resistance to our insecticides. Plant breeders keep ahead of these problems by crossbreeding with wild relatives of crop plants. Throughout history, over 7,000 kinds of plants have been used as food, and there may be another 75,000 species that are edible, with still others to be discovered. Increased salinization of soils may require that we switch, in parts of the world, to salt-tolerant plants such as Palmer saltgrass.

Tribal societies have long depended upon plants for curing ills, and today the pharmaceutical industry continually screens plants for useful drugs. In 1985 about 3,500 new chemical substances were discovered in natural sources. Most were obtained from higher plants, some from lower organisms (fungi, bacteria, and lichens), and some from marine organisms and various animals. Soil fungi, once considered insignificant, are now known to be important to the survival of many plants as well as a rich source of antibiotics. Vincristine, an extract from a Madagascar periwinkle plant, is now marketed for the control of childhood leukemia, while a close derivative is used for the treatment of Hodgkin's disease.

The need to cherish our forests is no better demonstrated than in the case of taxol, an extract of the bark of the Pacific yew, a tree discovered and described by Thomas Nuttall on his trip to the Far West. Taxol shows great promise as an anticancer drug, but many pounds of bark must be stripped to produce small quantities of the drug. Pacific forests are fast dwindling, and yews grow very slowly to a size that can produce appreciable amounts of taxol.

Cornell University biologist Thomas Eisner has spent much of his career at what he calls "chemical prospecting" among insects. Insects use an incredible array of substances in defense and communication; many of these have potential uses we are only beginning to appreciate. One of the substances in the blood of fireflies has recently been patented as an antiviral agent. A carrion beetle is known to produce a defensive chemical resembling progesterone, the agent in birth control pills. (Eisner asks, might it serve to reduce the birth rate in the rodents that are the beetles' major predators?) Animals of other groups also produce potentially useful chemicals. The African clawed frog exudes poisons from its skin that contain peptides that have suggested a cure for cystic fibrosis.

Eisner points out that while thousands of chemicals have been isolated from plants and animals and put to many uses, the majority remain to be discovered: "a vast, untapped treasury." As he says, different populations of any one species are likely to differ genetically as well as chemically, so it is not sufficient to study and to attempt to preserve a single population. This is especially true as we enter the era of biotechnology. The possibility of transferring genes from one organism to another, to produce greater viability, resistance to disease, or whatever, means that we cannot afford to lose genetic material of any kind through forced extinction of any species or population.

Sometimes the most improbable of animals have helped to solve problems of major interest. Neurophysiologists are finding how animals learn and remember by studying a sea snail, *Aplysia californica*, which has a relatively simple and accessible nervous system. Studies of the nervous system of the common squid of the Atlantic, *Loligo pealei* (named for Titian Peale's brother Rubens), have illuminated aspects of nerve transmission relevant to an understanding of Alzheimer's disease. Research on the enzymes that control the breakdown and regeneration of the claw muscles of molting Bermuda land crabs may help us to understand the causes of muscular dystrophy and similar diseases. Fruit flies of the genus *Drosophila* have taught us much of what we known of genetics (most species of *Drosophila* were not discovered until well into the twentieth century). The burgeoning science of molecular biology has depended heavily upon a lowly intestinal bacterium discovered in 1886 by the German bacteriologist Theodor Escherich and named for him: *Escherichia coli*. The same bacterium is a key species in tests for water contamination by fecal matter. It has been said that for every problem concerning living things there is an organism ideal for its solution. It is probable that there are still undiscovered species that hold the answers to problems that face us now or will in the future.

Terry Erwin of the Smithsonian Institution has recently found a wholly unexplored fauna in the tropical rain forest canopy. It is difficult to climb into the canopy, which may be a hundred feet or more above the ground through a tangle of formidable vines. Erwin and his colleagues shoot a projectile into the canopy with a line that catches on one of the upper branches. They then raise a canister that, on command, releases a fog of fast-acting insecticide. Insects and other small animals are then collected on sheets on the ground. From a single Panamanian tree species, *Luehea seemannii*, Erwin has collected over 1,200 species of beetles (and many other insects). About 13 percent of the beetles are believed to be host-specific; that is, they are associated only with one kind of tree. But the forests are filled with many other tree species, and Erwin estimates that in one hectare (2.5 acres) there may be 12,448 beetle species in the canopy, and perhaps over 30,000 species of insects of all kinds, the majority undescribed. Many of these are parasites or predators on other insects. Erwin asks: "Are there not among these some unknown beneficial treasures to be found? Do we not need to know, in these days of genetic engineering, what the world holds in the way of genetic diversity and genetic variability?"

In an address to the British Mycological Society in 1990, David L. Hawksworth presented data that suggest that there may be about a million and a half species of fungi in the world, though to date only about 69,000 have been described. Soil fungi associated with plant roots (mycorrhizas) are fundamental to the growth of at least 75 percent of the species of higher plants; other fungi assist in the breakdown and recycling of the tissues of dead organisms; still others serve as parasites in the natural control of other organisms. Commercial use of fungi includes the production of antibiotics, beer and wine, cheeses, breads, vitamins, and pesticides; and of course mushrooms are grown for direct consumption. The study of fungi, Hawksworth concludes, is "not for the faint-hearted, but a pleasure-ground for those seeking intellectually rewarding and also relevant endeavour."

Clearly there are many tasks ahead for naturalists, for in fact the delineation of species—though still far from complete—is only the first step in knowledge. We need to know what role each plays in nature, how it maintains its numbers in the face of competition and environmental change, how it may serve us in our own complex society, what it may teach us about ourselves. Those who speak of naturalists as "old-fashioned" fail to recognize that knowledge of biological diversity is basic to our continuing to live meaningful lives on our planet.

"Why should we care?" asks Edward Wilson in his book *The Diversity of Life*. "Let me count the ways." After reviewing some of the riches we derive from nature—foods, medicines, timber, and many other things—he goes on:

In amnesiac revery it is also easy to overlook the services that ecosystems provide humanity. They enrich the soil and create the very air we breathe. Without these amenities, the remaining tenure of the human race would be nasty and brief. The life-sustaining matrix is built of green plants with legions of microorganisms and mostly small, obscure animals—in other words, weeds and bugs. Such organisms support the world with efficiency because they are so diverse, allowing them to divide labor and swarm over every square meter of the earth's surface. They run the world precisely as we would wish it to be run, because humanity evolved within living communities and our bodily functions are finely adjusted to the idiosyncratic environment already created. . . . To disregard the diversity of life is to risk catapulting ourselves into an alien environment.

There is an important place for the nature writer, too, for it is he or she who speaks for nature on a more personal level and reminds us of its value in refreshing the human spirit. Good nature writing flourishes today as never before, and it is poignant as never before, as forests, plains, and even the oceans succumb to the pressures of a growing human population. The loss of species is proceeding at an alarming rate: according to Edward Wilson, perhaps 27,000 species are lost each year, 74 each day, 3 each hour. As Charles Bowden wrote in *Blue Desert*, with bleak irony: "For people who hate to learn the names of things, the world is getting better every day."

Peter Raven and Edward Wilson, in the article in *Science* cited earlier, have proposed a global biodiversity survey aimed at the identification of all species in order to plan for their conservation and possible use. They suggest a fifty-year program which would involve a great many more than the few hundred systematists now active. Such a program, they say, had best be done in the form of national biological surveys throughout the world, so that people of each country will understand and profit by their own country's biodiversity. The funding of such a project is not discussed. Somehow naturalists must learn to stimulate popular attention in the way NASA did a generation ago; they have much more to offer than the vacuities of space. The earth is our home; we cannot let so many of its remarkable living things disappear before we acknowledge their existence.

Perhaps, after all, we can build on NASA's accomplishments. Every manned venture into space, including the proposed space station, requires carrying earthly life-support systems—a lesson in our dependence upon them. And the view of earth from space, spinning brown, blue, and white in the vast emptiness, forever reminds us of how lucky we are to have been born into so sweet a home, a home replete with living things—a myriad of plants that produce our basic foods, provide the oxygen we breathe, and recycle our carbon dioxide; a multitude of animals that lend excitement to our forests and plains; and an underpinning of unseen things, the microorganisms that support life in so many ways. All of this on a ball of matter born of cosmic events billions of years ago and evolved from molecules in a primordial soup through events we can only vaguely fathom. Simply to gaze upon a meadow of flowers, to listen to a meadowlark, to smell the rich soil, should fill us with wonder and lead us to ask why we have not loved nature more.

Chronology of the Naturalists Discussed, By Year of Birth

(Major figures appear in bold)

1560–1621	Thomas Hariot
1650–1692	John Banister
1660–1740	Olaf Rudbeck
?1670–1711	John Lawson
?1679–1749	Mark Catesby
1694–1773	John Clayton
1699–1777	John Bartram
1707–1778	**Carl Linnaeus**
1709–1746	Georg Wilhelm Steller
1716–1779	Peter Kalm
1724–1766	Jane Colden
1729–1798	Johann Reinhold Forster
1739–1823	William Bartram
1743–1826	Thomas Jefferson
1746–1802	André Michaux
1751–1840	John Abbot
1753–1815	Gotthilf Heinrich Ernst Muhlenberg
1753–1828	Thomas Bewick
1754–1842	Archibald Menzies
1764–1848	John Barrow
1766–1813	**Alexander Wilson**

1766–1815   Benjamin Smith Barton
1769–1828   DeWitt Clinton
1770–1838   William Clark
1774–1809   Meriwether Lewis
1774–1820   Frederick Pursh
1775–?1849 William Bullock
1776–1842   Amos Eaton
1781–1862   Thomas Stewart Traill
1781–1866   George Ord
1782–1867   Alexander Philipp Maximilian
1783–1840   Constantine Samuel Rafinesque
1785–1851   **John James Audubon**
1786–1847   John Franklin
1786–1859   Thomas Nuttall
1787–1865   John Richardson
1787–1834   Thomas Say
1789–1855   William Swainson
1790–1836   William Leach
1790–1874   John Bachman
?1790–1835 Thomas Drummond
1793–1831   Johann Friedrich Eschscholtz
1793–1874   Gideon Lincicum
1793–1877   Jared Potter Kirtland
1793–1884   Almira Hart Lincoln
1794–1849   René Primevère Lesson
1794–1871   John Edwards Holbrook
1796–1843   Richard Harlan
1796–1852   William MacGillivray
1796–1861   John Stevens Henslow
1796–1873   **John Torrey**
1797–1861   Edwin James
1798–1834   David Douglas
1798–1864   William C. Cooper
1799–1863   Edward Harris

1799–1885   Titian Ramsay Peale
1800–1862   James Clark Ross
1801–1879   Ferdinand Jakob Lindheimer
1803–1857   Charles Lucien Bonaparte
1806–1850   Josiah Gregg
1806–1895   George Newbold Lawrence
1807–1873   Louis Agassiz
1809–1851   John Kirk Townsend
1809–1882   **Charles Darwin**
1809–1884   George Engelmann
1810–1888   **Asa Gray**
1810–1889   Frederick Adolph Wislizenus
1812–1853   John W. Gunnison
1812–1889   John Graham Bell
1813–1869   John Cassin
1813–1883   August Fendler
1813–1890   John Charles Frémont
1814–1880   Thomas Mayo Brewer
1815–1879   John P. McCown
1817–1886   Edward Tuckerman
1821–1849   William Gambel
1821–1904   Samuel Washington Woodhouse
1821–1912   Graceanna Lewis
1822–1893   George Vasey
1822–1895   Charles Girard
1822–1897   Darius Nash Couch
1822–1909   William Henry Edwards
1823–1887   **Spencer Fullerton Baird**
1823–1890   Charles Parry
1824–1911   William Wallace Anderson
1825–1883   John Lawrence LeConte
1825–1894   John Xántus
1826–1892   Sereno Watson
1827–1865   Adolphus Lewis Heermann

1827–1874 Bernard Rogan Ross
1831–1881 Martha Maxwell
1831–1911 Edward Palmer
1834–1895 Daniel Cady Eaton
1835–1866 Robert Kennicott
1835–1913 Philip Reese Uhler
1836–1897 Charles Bendire
1840–1897 Edward Drinker Cope
1840–1929 Henry Crècy Yarrow
1841–1930 Charles Andrew Allen
1842–1899 Elliott Coues
1843–1915 Edward Lee Greene
1845–1927 William H. Dall
1847–1879 Genevieve Estelle Jones
1849–1931 John Henry Comstock
1849–1934 Cornelia Clapp
1850–1929 Robert Ridgway
1850–1930 Henry Wetherbee Henshaw
1853–1907 Lucien Marcus Underwood
1854–1930 Anna Botsford Comstock
1855–1934 Edward William Nelson
1855–1942 **C. Hart Merriam**
1856–1916 Edgar Alexander Mearns
1857–1925 Thomas Lincoln Casey
1858–1914 Frederick True
1858–1934 Elizabeth Britton
1859–1934 Nathaniel Lord Britton
1859–1952 Aven Nelson
1859–1953 Alice Eastwood
1860–1931 Per Axel Rydberg
1868–1953 Nathan Banks
1873–1946 Edward Alfonso Goldman
1884–1941 Alexandre Arsène Girault

References

GENERAL

Anderson, Berta. 1976. *Wild Flower Name Tales.* Colorado Springs, Colo.: Century One Press.

Choate, E. A. 1985. A *Dictionary of American Bird Names.* Revised by R. A. Paynter, Jr. Boston: Harvard Common Press.

Dictionary of American Biography. 1928–1981. 20 vols., 8 suppl. New York: Scribner's.

Dictionary of National Biography: From the Earliest Times to 1900. 1921–1922. 22 vols. Edited by L. Stephen and S. Lee. London: Oxford Univ. Press.

Elman, Robert. 1982. *America's Pioneering Naturalists.* Tulsa, Okla.: Winchester Press.

Ewan, Joseph. 1950. *Rocky Mountain Naturalists.* Denver: Univ. Denver Press.

Ewan, Joseph, ed. 1969. A *Short History of Botany in the United States.* New York: Hafner Publ. Co.

Geiser, S. W. 1948. *Naturalists of the Frontier.* Dallas: University Press in Dallas.

Goode, G. B. 1991. *The Origins of Natural Science in America: The Essays of George Brown Goode.* Edited by Sally G. Kohlstedt. Washington: Smithsonian Institution Press.

Harshberger, J. W. 1899. *The Botanists of Philadelphia and their Work.* Philadelphia: T. C. Davis & Sons.

Hume, Edgar E. 1942. *Ornithologists of the United States Army Medical Corps: Thirty-six Biographies.* Baltimore: Johns Hopkins Press.

Kastner, Joseph. 1977. A *Species of Eternity.* New York: Alfred A. Knopf.

McKelvey, Susan D. 1955. *Botanical Exploration of the Trans-Mississippi West 1790–1850.* Jamaica Plain, Mass.: Arnold Arboretum of Harvard Univ.

Mallis, Arnold. 1971. *American Entomologists*. New Brunswick, N.J.: Rutgers Univ. Press.

Mearns, Barbara, and Richard Mearns. 1988. *Biographies for Birdwatchers: The Lives of Those Commemorated in Western Palearctic Bird Names*. London: Academic Press.

Meisel, Max. 1924–1929. A *Bibliography of American Natural History. The Pioneer Century, 1769–1865*. 3 vols. Brooklyn, N.Y.: Premier Publ. Co.

Palmer, T. S. 1928. Notes on persons whose names appear in the nomenclature of California birds. *The Condor*, 30: 261–307.

Peattie, D. C. 1938. *Green Laurels: The Lives and Achievements of the Great Naturalists*. New York: Garden City Publ. Co.

Stearns, R. P. 1970. *Science in the British Colonies of America*. Urbana: Univ. Illinois Press.

Stresemann, Erwin. 1975. *Ornithology from Aristotle to the Present*. Translated by H. J. and C. Epstein. Cambridge, Mass.: Harvard Univ. Press.

A

Dow, R. P. 1914. John Abbot, of Georgia. *Journal of the New York Entomological Society*, 22: 65–72.

Heinrich, Bernd. 1981. *Insect Thermoregulation*. New York: John Wiley and Sons.

Knight, D. M. 1986. William Swainson: Naturalist, author and illustrator. *Archives of Natural History*, 13: 275–290.

Largen, M. J., and V. Rogers-Price. 1985. John Abbot, an early naturalist-artist in North America: His contributions to ornithology. . . . *Archives of Natural History*, 12: 231–252.

Remington, C. L. 1948. Brief biographies 10: John Abbot (1751–183-). *Lepidopterist's News*, 2: 28–30.

B

Cutright, P. R., and M. J. Brodhead. 1981. *Elliott Coues: Naturalist and Frontier Historian*. Urbana: Univ. Illinois Press.

Dall, W. H. 1915. *Spencer Fullerton Baird: A Biography*. Philadelphia: Lippincott.

Deiss, W. A. 1980. Spencer F. Baird and his collectors. *Journal of the Society for the Bibliography of Natural History*, 9: 635–645.

Drury, W. H., Jr. 1961. The breeding biology of shorebirds on Bylot Island, Northwest Territories, Canada. *The Auk*, 78:176–219.

Rivinus, E. F., and E. M. Youssef. 1992. *Spencer Baird of the Smithsonian*. Washington: Smithsonian Institution Press.

C

Cutright, P. R. 1969. *Lewis and Clark: Pioneering Naturalists.* Urbana: Univ. Illinois Press.

Lewis, H., and P. H. Raven. 1958. Rapid evolution in *Clarkia. Evolution,* 12: 319–335.

D

Davies, John. 1980. *Douglas of the Forests: The North American Journals of David Douglas.* Seattle: Univ. Washington Press.

Herman, L. M., ed. 1980. *Cetacean Behavior: Mechanisms and Functions.* New York: John Wiley and Sons.

Merriam, C. H. 1927. William Healey Dall. *Science,* 65: 345–347.

E

Peattie, D. C. 1950. A *Natural History of Western Trees.* Cambridge, Mass.: Riverside Press.

F

Graustein, J. 1967. *Thomas Nuttall, Naturalist: Explorations in America 1808–1841.* Cambridge, Mass.: Harvard Univ. Press.

Hoare, M. E. 1971. Johann Reinhold Forster (1729–1798): Problems and sources of biography. *Journal of the Society for the Bibliography of Natural History,* 6: 1–8.

G

Bailey, F. M. 1928. *Birds of New Mexico.* Santa Fe: New Mexico Department of Fish and Game.

Bent, A. C. 1932. Life Histories of North American Gallinaceous Birds. *Bulletin of the U.S. National Museum,* 162: 73–83.

H

Darwin, Charles. [1892] 1958. *The Autobiography of Charles Darwin.* Edited by Francis Darwin. New York: Dover Publ. Co.

Ford, Alice. 1988. *John James Audubon: A Biography.* New York: Abbeville Press.

I

Gordh, G., A. S. Menke, E. C. Dahms, and J. C. Hall. 1979. *The Privately Published Papers of* A. A. *Girault.* American Entomological Institute, Memoir 28.

J

Benson, Maxine, ed. 1988. *From Pittsburgh to the Rocky Mountains: Major Stephen Long's Expedition 1819–1820.* Golden, Colo.: Fulchrum.

Dupree, A. Hunter. 1959. *Asa Gray 1810–1888.* Cambridge, Mass.: Harvard Univ. Press.

Lurie, Edward. 1960. *Louis Agassiz: A Life in Science.* Chicago: Univ. Chicago Press.

K

Wright, A. H., and A. A. Wright. 1957. *Handbook of Snakes.* 2 vols. Ithaca, N.Y.: Cornell Univ. Press.

L

Essig, E. O. 1931. John L. LeConte. In A *History of Entomology.* New York: Macmillan.

Merriam, C. H. 1895. The LeConte thrasher, *Harporhynchus lecontei. The Auk,* 12: 54–59.

M

Galloway, D. J., and E. W. Groves. 1987. Archibald Menzies MD, FLS (1754–1842), aspects of his life, travels and collections. *Archives of Natural History,* 14: 3–43.

King, John A., ed. 1968. *Biology of* Peromyscus (*Rodentia*). American Society of Mammalogists, Special Publ. 2.

Richmond, C. W. 1918. In memoriam: Edgar Alexander Mearns. *The Auk,* 35: 1–18.

Sterling, K. B. 1974. *Last of the Naturalists: The Career of C. Hart Merriam.* New York: Arno Press.

N

Finley, R. B., Jr. 1958. *The Wood Rats of Colorado: Distribution and Ecology.* Univ. Kansas Publ., Museum of Natural History, 10: 213–552.

Goldman, E. A. 1935. Edward William Nelson—Naturalist, 1855–1934. *The Auk,* 52: 135–148.

Linsdale, J. M., and L. P. Tevis, Jr. 1951. *The Dusky-footed Wood Rat.* Berkeley: Univ. California Press.

Williams, R. L. 1984. *Aven Nelson of Wyoming.* Boulder: Colorado Associated Univ. Press.

Young, S. P. 1947. Edward Alphonso Goldman: 1873–1946. *Journal of Mammalogy,* 28: 91–102.

O

Cantwell, Robert. 1961. *Alexander Wilson: Naturalist and Pioneer, a Biography.* Philadelphia: Lippincott.

Janzen, D. H. 1977. Why fruits rot, seeds mold, and meat spoils. *American Naturalist,* 111: 691–713.

Reichmann, O. J., and C. Rebar. 1985. Seed preferences by desert rodents based on levels of mouldiness. *Animal Behaviour,* 33: 726–729.

P

McVaugh, Rogers. 1956. *Edward Palmer: Plant Explorer of the American West.* Norman: Univ. Oklahoma Press.

Yensen, N. P., and S. B. de Yensen. 1987. Development of a rare halophytic grain: Prospects for reclamation of salt-ruined lands. *Journal of the Washington Academy of Sciences,* 77: 209–214.

Q

Ferris, C. D. 1973. A revision of the *Colias alexandra* complex (Pieridae) aided by ultraviolet reflectance photography. . . . *Journal of the Lepidopterist's Society,* 27: 57–73.

Hibbert, C. 1976. *The Royal Victorians: King Edward VII, His Family and Friends.* Philadelphia: Lippincott.

Silberglied, R. E., and O. R. Taylor, Jr. 1978. Ultraviolet reflection and its behavioral role in the courtship of the sulphur butterflies. . . . *Behavioral Ecology and Sociobiology,* 3: 203–243.

R

Hill, J. E., and J. D. Smith. 1984. *Bats: A Natural History.* Austin: Univ. Texas Press.

Rafinesque, C. S. 1978. *Autobiography,* with Biographies and Bibliographies by R. E. Call and T. J. Fitzpatrick. New York: Arno Press.

S

Bent, A. C. 1942. *Life Histories of North American Flycatchers, Larks, Swallows, and Their Allies.* Bulletin of the U.S. National Museum, no. 179.

Stroud, P. T. 1992. *Thomas Say: New World Naturalist.* Philadelphia: Univ. Pennsylvania Press.

Viola, H. J., and C. Margolis, eds. 1985. *Magnificent Voyages: The U.S. Exploring Expedition, 1838–1842.* Washington: Smithsonian Institution Press.

Weiss, H. B., and G. M. Ziegler. 1978. *Thomas Say: Early American Naturalist.* New York: Arno Press.

T

Gorman, M. L., and R. D. Stone. 1990. *The Natural History of Moles.* Ithaca, N.Y.: Cornell Univ. Press.

Mellanby, Kenneth. 1971. *The Mole.* London: Collins.

Townsend, John Kirk. [1839] 1978. *Across the Rockies to the Columbia.* Lincoln: Univ. Nebraska Press.

U

Carpenter, F. M., and P. J. Darlington, Jr. 1954. Nathan Banks: A biographical sketch and list of publications. *Psyche,* 61: 81–110.

Robinson, M. H. 1981. A stick is a stick and not worth eating: On the definition of mimicry. *Biological Journal of the Linnean Society,* 16: 15–20.

Schwarz, E. A., O. Heidemann, and N. Banks. 1914. Life and writings of Philip Reese Uhler. *Proceedings of the Entomological Society of Washington,* 16: 1–7.

V

Griscom, L., and A. Sprunt, Jr. 1957. *The Warblers of America.* New York: Devin-Adair.

Mead, Chris. 1983. *Bird Migration.* New York: Facts on File.

W

Stone, Witmer. 1905. News and Notes: Dr. Samuel W. Woodhouse. *The Auk,* 22: 104–106.

Sullivan, B. K. 1982–1983. Sexual selection in Woodhouse's toad (*Bufo woodhousei*). *Animal Behaviour,* 30: 680–686; 31: 1011–1017.

Sullivan, B. K. 1985. Male calling behavior in response to playback of conspecific advertisement calls in two bufonids. *Journal of Herpetology,* 19: 78–83.

Wislizenus, F. A. [1912] 1969. *A Journey to the Rocky Mountains in the Year 1839.* Glorieta, New Mexico: Rio Grande Press.

X

Xántus, John. [1858] 1975. *Letters from North America.* Translated and edited by T. Schoenman and H. B. Schoenman. Detroit: Wayne State Univ. Press.

Xántus, John. [1858] 1976. *Travels in Southern California.* Translated and edited by T. Schoenman and H. B. Schoenman. Detroit: Wayne State Univ. Press.

Zwinger, Ann. 1986. *John Xántus: The Fort Tejon Letters 1857–1859.* Tucson: Univ. Arizona Press.

Y

Hunsaker, Don, II. 1962. Ethological isolating mechanisms in the *Sceloporus torquatus* group of lizards. *Evolution,* 16: 62–74.

Lanham, Url. 1973. *The Bone Hunters.* New York: Columbia Univ. Press.

Noble, G. K. 1934. Experimenting with the courtship of lizards. *Natural History,* 34(1): 5–15.

Smith, H. M. 1946. *Handbook of Lizards: Lizards of the United States and Canada.* Ithaca, N.Y.: Cornell Univ. Press.

Z

Goodwin, Derek. 1967. *Pigeons and Doves of the World.* Ithaca, N.Y.: Cornell Univ. Press.

WOMEN IN NATURAL HISTORY
OF THE NINETEENTH AND EARLY TWENTIETH
CENTURIES

Benson, Maxine. 1986. *Martha Maxwell: Rocky Mountain Naturalist*. Lincoln: Univ. Nebraska Press.

Bonta, Marcia Myers. 1991. *Women in the Field: America's Pioneering Women Naturalists*. College Station: Texas A&M University Press.

Comstock, A. B. 1953. *The Comstocks of Cornell: John Henry Comstock and Anna Botsford Comstock*. Ithaca, N.Y.: Cornell Univ. Press.

Rossiter, Margaret. 1982. *Women Scientists in America: Struggle and Strategies to 1940*. Baltimore, Maryland: Johns Hopkins Univ. Press.

Wystrach, V. P. 1977. Anna Blackburne (1726–1793)—A neglected patroness of natural history. *Journal of the Society for the Bibliography of Natural History*, 8: 148–168.

NATURALISTS, THEN AND NOW

Bates, Marston. 1950. *The Nature of Natural History*. New York: Scribner's.

Eisner, Thomas. 1992. The hidden value of species diversity. *Bioscience*, 42: 578.

Erwin, T. L. 1983. Tropical forest canopies: The last biotic frontier. *Bulletin of the Entomological Society of America*, 29: 14–19.

Hawksworth, D. L. 1991. The fungal dimensions of biodiversity: Magnitude, significance, and conservation. *Mycological Research*, 95: 641–655.

Raven, P. N., and E. O. Wilson. 1992. A fifty-year plan for biodiversity surveys. *Science*, 258: 1099–1100.

Wilson, E. O., ed. 1988. *Biodiversity*. Washington: National Academy of Sciences Press.

Wilson, E. O. 1992. *The Diversity of Life*. Cambridge, Mass.: Harvard University Press.

Index